FREE-RANGE
chicken gardens

FREE-RANGE
chicken gardens
How to Create a Beautiful, Chicken-Friendly Yard

Jessi Bloom

WITH PHOTOGRAPHS BY
Kate Baldwin

Timber Press
Portland · London

To my grandmother, **Iva**
And all who welcome chickens into their gardens

table of contents

introduction

WHEN I FIRST GOT CHICKENS, I MADE A LOT OF MISTAKES. In the first few weeks of letting them roam freely in the garden, you'd often find me chasing them around in circles, trying to get them to go where I wanted, which is about as easy as herding cats. Our first chicken coop provided adequate shelter, but ended up being better housing for rats than birds. And I wanted to collect breeds just like I collect plants—at least one of every kind, please—ignoring my husband's warnings of becoming a chicken hoarder. Fast forward, and now our girls will come when called and even hop inside visitors' cars as if they are ready to hit the town. Their housing is clean, rodent-proof, and an impenetrable barricade from night predators.

Their days are spent weeding the garden beds, mowing the lawn, and chasing insects, between regular dust baths to groom themselves and lounging in the sun. It's a great arrangement for everyone.

Chickens are easily one of the most useful animals we can have in our lives. Unlike other pets we keep, chickens provide us with food—fresh from our own backyard. For gardeners, chickens can be a resourceful tool as well as a companion, but there is much to know so the birds don't wreak havoc in your garden.

There are not many resources available for gardeners who would like to know what's involved in keeping chickens, or for chicken owners who want a beautiful garden with free-ranging hens in it. This book takes you through the basics of starting with chickens, from how many to get, to what breeds will be best for you, to acclimating them to your garden and routines. It also covers the essentials of keeping your chickens protected, even training the birds. The heart of the book has you looking at your garden as habitat for your flock, starting with the basic elements of landscape design, then selecting materials for fencing and hardscapes, onto choosing chicken-friendly plants, and reviewing sample garden designs. One chapter is devoted to innovative coop design, followed by profiles of predators and information on health care. Each dimension of the book is explored with photos and illustrations.

While doing research for the book, I interviewed gardeners with chickens from throughout the country, and I found that they all did things differently. Their gardens varied, and they faced different challenges. In each chapter, I've included a story on a chicken keeper who has a successful system of chickens in the garden. It's my hope that this book will inspire gardeners to become chicken-keepers and create a partnership with their chickens in having a functional and beautiful outdoor space.

CHICKENS AND GARDENS
working together

With food politics currently in the media forefront and self-sufficiency becoming mainstream, victory gardens are being built again everywhere, and across the country citizens are banding together to legalize poultry in their backyards. More and more, chickens are becoming part of our gardens, providing us with fresh eggs, but their strengths as garden helpers are often overlooked.

jessi's girls

I **LIVE ON A SMALL PIECE OF LAND NORTH OF SEATTLE,** Washington, with my husband and two young boys, amidst lush gardens, dogs, ducks, turkeys, a goat, a horse, and about a dozen chickens who roam freely, greeting visitors and tending the gardens. The girls free range during the day and roost at night in a small barn with other animals. The roof water from the building is collected in a 300-gallon cistern and overflows into a trough, then into a small pond. We've kept all different breeds of chickens and have offered sanctuary and a home to many rescued animals. The chickens are our "pets with benefits": they provide food, fertilizer, and garden help, and they also teach my children lessons about where their food comes from and about the responsibility of caring for the birds.

Our gardens are completely organic, with a mixture of native plants, ornamentals, and edibles planted together in different beds, layers, and garden rooms. Our chickens rarely eat plants that are not meant for them to eat, and have never had damaging levels of parasites or serious disease. We have very few predator problems because of the coverage from plants, fencing, and protective animals we keep such as dogs. We've always let our chickens free range, and we have adapted by protecting particular plants when necessary and arranging for the birds to help us with garden chores by using a chicken tractor, which is a bottomless portable pen.

The chickens offer a sense of humility and peace, which helps keep me grounded. If I'm having a bad day, I only need to spend a few minutes with them before I feel good again. Their silly antics make us all laugh, and their sense of family is inspiring to watch. They have complex social lives and distinct personalities, and they really do take care of each other but still have occasional tiffs. I couldn't imagine having a garden without chickens.

TOP: A Rhode Island red hen walks up a garden path. BOTTOM RIGHT TO LEFT: I am sitting in the garden with one of my favorite gardening buddies, an Easter Egg hen. This black hen can find shelter in the plants along the driveway if a predator flies overhead.

CHICKENS ARE TERRIFIC GARDENING ASSISTANTS with natural soil-building capabilities, and they help to manage pests and weeds. Much like other pets we keep, they are easy to care for, can be trained to come when called or to do tricks, and some people even bring them inside their homes much like other domesticated birds. Owning chickens has become appealing to a wide audience, from young families who are homesteading to baby boomers with empty nests who are looking for new hobbies and interests. Seasoned gardeners are now looking into getting chickens and wondering how to take full advantage of their benefits while protecting their gardens and hard-earned crops.

Chickens have long been an integral part of human society and the food chain. Easily the most useful animal we ever domesticated, chickens require very little care and land compared to other livestock. They are one of the oldest domesticated animals, originating in Asia, with history dating back at least 5000 years. The Wild Red Junglefowl (*Gallus gallus*), which looks similar to a Brown Leghorn but smaller, is thought to be the oldest ancestor of today's domestic chicken (*Gallus domesticus*).

Chickens are gallinaceous: they stay close to their home residence and roost there at night, unlike many common wild birds that migrate. This trait makes them ideal for living symbiotically with us and enables us to allow them to free range. Letting them live as they would naturally is good for their health and well-being, but needs some planning.

 ### INCORPORATING CHICKENS INTO YOUR GARDEN

Our gardens can be great chicken habitats if designed well. Plants provide many benefits for us, including food, shelter, shade, and aesthetic pleasure. Chickens need a diverse plant community to live in, with layers of plants to hide among and browse from. If your garden has only a few plants, chickens can wreak havoc on those plants because of hunger or boredom. So the more options you give them in your garden, the less damage they will do. Other factors that will contribute to your success with free-range chickens include the number of chickens you have, the size of your garden, and how you manage your garden.

Chickens can provide us with many benefits:

o *Garden help:* If you design your garden system to employ chickens, they can keep weeds down, keep pest populations in check, help aerate the soil, help compost greens and food waste, and even help mow your lawn.

o *Fertilizer:* Chicken manure is a resource for rich, soil-building fertilizer.

o *Food:* A homegrown protein source, chicken eggs and meat are nutritious, and you can't beat harvesting fresh eggs from your backyard.

Let's define what we mean by "garden." As a landscape designer, I find that people use the words "landscape" and "garden" interchangeably. Many people gasp at the thought of letting chickens into their garden, visualizing a carefully tended vegetable bed or patch of delicate annuals growing in a special area of the yard. Generally speaking, most concepts of a garden include an area of cultivated soil for growing plants. In my mind, a garden means your entire backyard or property, and in this case with chickens.

When I told chicken owners I was writing a book about gardening with chickens, many often looked at me like I was crazy. I've never kept my chickens confined during the day, and I've used creative solutions for the instances when my chickens and my garden didn't mix well. It takes planning, some patience, and time, but that is what gardening entails anyway, with or without chickens.

OPPOSITE, TOP: My chickens forage near a vegetable garden.
OPPOSITE, BOTTOM RIGHT TO LEFT: A Brown Leghorn peeks out from an oakleaf hydrangea. A hen walks in the understory of a forest garden.

Many years ago at a convention for green living, in my booth I had a large photograph of three of my hens foraging in the grass in front of my vegetable garden. Several people who saw that photo insisted that it must have been staged. I had taken the photo to highlight the cobalt blue containers at the entrance of the garden, and the chickens happened to walk in front of the camera, and their vivid colors enhanced the photo of the garden, so I went with it. I've talked with many chicken owners, and discovered that at one time they let their chickens range, but after a devastating incident with a predator, they vowed never to let their hens be in harm's way again. Some were sick of cleaning up after the chickens when they were let out or didn't want their crops eaten and hard work ruined. For convenience and through fear, many people leave their birds locked up. Some people only let their birds out on a schedule that varies from season to season to avoid garden invasion.

Just as there are many ways to garden, there are many ways to keep chickens. Everyone has different methods of raising their birds, and there is no one correct way to do it. Even if you cannot let your chickens free range, this book can give you ideas on plants to grow for chickens to eat, creative housing with fenced runs, dealing with predators, and common health issues.

factors to consider

The important factors in having a successful free-range chicken garden are:

○ **THE GARDEN:** the size of your space, how the space is designed, the plants growing in it, and how it is managed and maintained.

○ **THE CHICKENS:** the number of birds in the flock, what breeds you choose, and how they are raised.

○ **THE ROUTINE:** the chickens' feeding schedule, and timing and access to the garden.

There are many ways to keep chickens on your property, all of which have their pros and cons. Proper land management is key, as is knowing what stocking rate your land can accommodate, which is how many animals to keep on a specific amount of land. Any animal we keep has the potential to graze the land to a detrimental extent if we don't plan and manage well for its presence. It's important to know your land—soil type, sun exposure, and such factors—to help you figure out what chicken-keeping method is best for you.

Free Range

"Free range" means different things throughout the world. In many cases, free range means that the animals are allowed to graze freely on the land outdoors. For commercially grown poultry in the United States, however, "free-range" chickens are commonly packed in large poultry houses and are given access to the outdoors through a small door leading to a small, enclosed exterior run. The USDA's definition of free range is broad: "Producers must demonstrate to the Agency that the poultry has been allowed access to the outside." Chickens that truly free range have full access to the outdoors, and can forage (find their own food) in the grass in the sunlight and fresh air. By allowing them access to fresh soil filled with organisms and a lush, healthy landscape filled with plants, the chickens can use the plants to hide in or eat from, which increases their quality of life and overall health.

Keeping a free-range garden means allowing the chickens access to the outdoors whether it is all the time or only on weekends. Even if laws or predators in your area prevent the chickens from having free-range access, there are still many ways chickens and gardens can work together for mutual benefit.

○ *Full-time free range.* This method leaves chickens out all day and night, which is closest to their natural state of living in the wild but is rarely done with most backyard flocks. Letting chickens roost in trees at night, as they would naturally do, heightens the risk of predator attack. Chickens living in a completely free-range way, fending for themselves, may not be as social with humans or tame enough to handle. This method means there's no coop to build or clean, but it also means it may not be easy to find the chickens' eggs.

○ *Part-time free range.* This method is how I would define my own chicken-keeping practice. The approach requires a garden that is designed and managed accordingly and it means that your birds must be acclimated to and trained for your routine. It means less coop cleaning, the hens lay their eggs in the coop, and you still have sociable birds. The chickens live much like they would in the wild, foraging for food during the day, which cuts down on feed costs and can help out in the garden. They have access to the coop during the day to lay their eggs, and are closed in the coop for the night to roost.

o *Occasional free range.* I think of this method as weekend free ranging. This approach is used by chicken keepers who are often not home during the day to watch over the flock or only want to let the chickens out when they have full supervision. It benefits the chickens by letting them scratch the earth for edibles such as insects and forage as they would naturally. The method is commonly used in areas where predator risk is high or gardens need more protection.

Confined Range

Chickens can be kept in confined range systems, which offer the behavioral and health benefits of free ranging, but also benefit the gardener by only allowing them to scratch, graze, and fertilize in targeted areas. These methods can also be utilized at certain times of the year when needed most.

o *Pastured.* This practice is common on big rural lots and is great for large numbers of birds if you have the room. Pastured ranging puts chickens to work in a particular area of the landscape by having them temporarily fenced in to graze, with a simple house or coop structure to keep them safe at night. The method is commonly used by small commercial poultry growers. The approach allows you to move the pasturing setup from one area to another after the chickens have grazed down the grass. The grass gets "mowed" by the chickens as they graze, and the land gets naturally fertilized by the chickens, building soil fertility over the area. The grass in a pasture area should be no more than 5 inches tall when the chickens are introduced, and the chickens should be removed when it's grazed down to about 2 inches. If rotation isn't timed carefully and the birds are left in one area too long, the ground can become barren of vegetation and the soil structure can be damaged from compaction. If there are no plants for shelter in the area being ranged by chickens, then housing should be large enough and up off the ground so the chickens can take cover in case of an overhead predator attack.

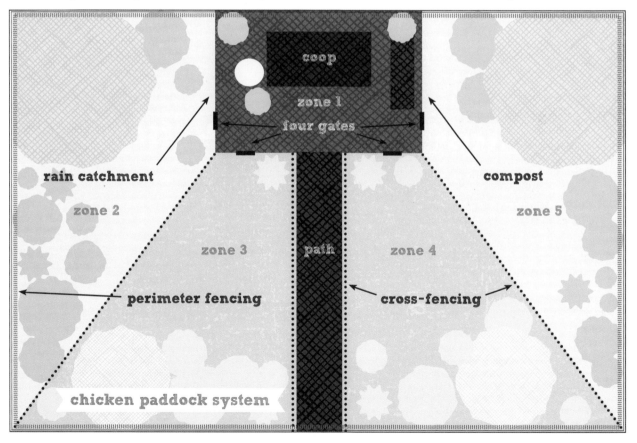

o *Paddocks.* Paddocks are usually fixed pens with perimeter fencing that allow chickens to be rotated to and from specific zones at certain times of the year. Chicken portals or small gates open from the central coop area, so chickens can forage in a particular paddock. This method combines many great features of the free-range methods and pasturing with cross-fencing. Each paddock contains different plant communities, and the chickens are allowed to forage and range in a particular paddock when it is needed or safe for the plants in that paddock. For instance, one paddock might have blueberry bushes and the chickens can range there most of the year except when the blueberries are ripening, or different greens or grains can be grown as the groundcover of an orchard for the chickens to graze, and then you move the chickens so the groundcover can recover. Each paddock also has plants for shelter and protection from predators.

o *Tractored.* Another common way to put chickens to work is to use a chicken tractor, a movable floorless pen. With this method, you can have the chickens work on the soil in small areas such as raised beds or on a backyard lawn. Chicken tractors can be all shapes and sizes. Tractors are perfect for home-scale chicken keepers, or farmers who need to get chickens to work in specific areas or at certain times of the year. In farm applications, chickens can be used to follow after cattle foraging an area; in areas where a large density of cow pies has accumulated, the chickens will glean undigested seeds and parasites, and will help spread the manure so it doesn't create spots where grass refuses to grow. Tractors can be permanent housing for chickens, but you need to have enough land to rotate to new grass or pasture as often as needed, which in turn relates to how many chickens you have and the size of the tractor.

tip

It is not practical to grow plants within the run of confined chickens. If they do not have choices in their foraging, they will eventually devour anything green.

Confined

Not all chickens are going to have access to free range or be put to work, whether there are laws prohibiting it in your area, high predator risk, or other hurdles. Chickens can have healthy, happy lives in a confined system, if certain issues are considered and accommodated.

o *Coop and run.* This method is a common way to keep chickens, especially in urban environments. This approach means there is less risk of predators coming after the chickens. But there is more coop cleaning, and if the space is not big enough and the birds are overcrowded, struggles can develop within the pecking order and low-ranking chickens can be bullied and stressed. Coops can be built in all sizes, shapes, and styles and attached to the coop is a large permanent run that can be covered or not. This method of keeping chickens will involve more feed costs because the chickens won't be able to forage their food. Even if you are not able to give the chickens access to free range, your birds can still be utilized with a tractor system.

o *Caged.* Most commercial laying hens live out their lives in battery cages, which sometimes aren't large enough to allow the birds to spread their wings or even turn around, leading to serious health problems. Many chicken owners keep their own flock to give the birds a better life, so caging is not a common method of containing them in backyards. Cages should be used only for transporting birds and for confinement in case of illness or emergency.

Chickens can be kept in any of these ways, and these methods can be used in conjunction with one another, at different times of the year as the growing season brings on plenty of food for the chickens or when vegetation is scarce. When you are first starting with chickens, you may not know which methods will work best for you. Over time, your goals may change, and you'll try something new. Initially deciding on a method will depend on your lifestyle, your garden, and your chickens.

OPPOSITE: A chicken tractor is utilized in a vegetable bed for weed control, pest control, and soil building.

CHICKENS AND SUSTAINABILITY

Over past decades, new generations of environmentally aware consumers have been looking for healthier alternatives to an industrial food system, where our food is grown in chemically dependent monocrop systems, animals are mistreated, and the safety of what we eat is often in question.

As we look for ways to become less dependent on industrial farming and more self-sufficient, many households have been turning to growing their own food. Chickens offer us food security, because they can provide us with an excellent source of homegrown protein. Chickens are making their way into many urban homesteads because they can be utilized for more than food, much like they are in other alternative farm systems, such as permaculture and biodynamics. These farm systems utilize ecologically sound design and resource management, such as building and maintaining proper soil health as a foundation for sustainability. Many design principles from these systems can be incorporated into our gardens.

Chickens and Permaculture

Permaculture is simply defined as "permanent agriculture," which encompasses many fields of study about how to maintain a self-sufficient lifestyle and an agriculturally productive ecosystem. Developed in the 1970s by Australians Bill Mollison and David Holmgren, this design system aims at being sustainable, while producing maximum yields with minimum work because all elements in the system are designed to work together and benefit each other. Chickens are a classic example of a working element in permaculture design. In considering their input and output in the system, the needs of chickens are simple: Chickens need food, water, and shelter, and in return, they provide manure, which could be considered waste but is actually a great resource of soil nutrients. Their eggs and meat are food, and their natural behaviors such as scratching (tilling the soil) and insect eating (pest control) can be utilized as work we would otherwise need to do.

OPPOSITE, TOP: The Bullock Permaculture Homestead, Orcas Island, Washington, keeps chickens in a pasture system during the summer in a young plum orchard. BOTTOM LEFT TO RIGHT: The main vegetable garden at S&S Homestead on Lopez Island, Washington, is fenced for protection from deer and chickens, who patrol the perimeter in search of insects. A hen uses a portal tunnel between the coop and paddock.

Chickens and Biodynamics

Biodynamic farming was first conceived in 1924 by Rudolf Steiner in Germany. This holistic approach treats the farm as an integrated, self-sufficient organism. The health of the soil is maintained with nutrient recycling, whether it is organic preparations from manures or plants, which then in turn feeds the plants, animals, and people of the farm.

I recently visited an inspiring biodynamic farm, the S&S Homestead Biodynamic Farm, on Lopez Island, Washington. There, chickens are used as pest control both while free ranging and in tractors over areas of accumulated cow manure. On this biodynamic farm, nothing goes to waste. Even chicken feathers, rich in nitrogen, are used to line seedling trays. Part of their land is donated space for CSA (Community Supported Agriculture), where people without land can grow their own food. Their farm is also a part of the local school district's curriculum in which all students K through 12 spend time learning about farming.

CHICKENS AND YOUR LIFESTYLE

If you have had pets and children, then you are probably pretty familiar with being responsible for another life and the art of handling feces without being squeamish. These are two qualities a chicken keeper must possess. Fortunately, chickens are much easier to raise than children and easier to keep than many other common pets like dogs. But keeping chickens does involve some work, especially the initial setup, and they need to be fed, given fresh water daily, and cleaned up after regularly.

Chickens are versatile creatures and can live harmoniously with people from all kinds of backgrounds and living situations. They get along really well with other animals, providing that the other animals don't want to eat them, and some domesticated pets can even be trained to protect your flock. Chickens can be a wonderful educational tool and companion for children, and are quickly replacing the hamster and other childhood pets.

If you work long, irregular hours and travel for long periods of time, you need to have a plan in place to make sure the birds are taken care of. Some neighbors might gladly check on your chickens while you are away in exchange for some fresh eggs.

Your Gardener Personality

What kind of gardener are you? Are you a collector of rare plants that are irreplaceable? Are you patient and laid back, understanding that life in the garden is an ecological process that takes time and is constantly evolving? Could you care less about plants and simply use your backyard to entertain friends and family? Gardener personalities vary, and not all gardeners are going to get along with free-ranging chickens.

In my garden I have spent years changing the garden layout, adding plants, moving plants, nursing plants, pruning plants. I find this endless process of creative play therapeutic. On any given day, I can come home to find mulch scratched out of the bed, or a few new piles of manure on the pathway—often when I expect guests—but I can quickly clean up after the birds if I need to. They have taught me patience with humor in the garden. But if your garden must be perfect, and you can't stand plants occasionally being picked at, then you might not want chickens free ranging unless you closely supervise them.

Choosing Your Flock

Chickens are relatively easy on the land compared to other livestock, thus making them an ideal animal to keep for almost anyone. There are many factors to consider, though, when selecting chickens for your flock.

The question of how many chickens to keep in your yard is a crucial one. With chickens, because there are so many ways to raise them and so many different kinds of garden settings, there is no straight answer, other than using practical sense, for how many to keep on a certain amount of land unless the law restricts the amount. If you are going to raise the birds for eggs for your own family's use, then think about how many eggs you can eat in one week. A good rule of thumb is to have two birds for each person who eats eggs in your household. One hen can lay anywhere from 50 to 300 eggs per year, depending on breed, age, and environmental factors. Keeping chickens at different ages will ensure more consistent egg production, because when chickens get older, their egg laying slows down.

number of chickens for your garden		
SIZE OF GARDEN	**NUMBER OF CHICKENS**	**CHICKEN SYSTEM**
Small urban lot (under 7000 sq ft total lot including house and garage)	3–5 chickens. The smaller the space, the fewer or smaller the birds; consider bantams	Coop and run, paddocks, part-time or occasional free-range schedule with seasonal rotation and tractoring
Suburban lot (7000–13,000 sq ft)	5–8 chickens	Any of the methods
Large suburban lot or rural (over 13,000 sq ft)	8–12 chickens or more	Any of the methods

the chicken life cycle

If you are going to keep chickens, it is important to understand their life cycle. Chickens have a relatively short life span, potentially seven or more years. Hens usually stop laying eggs after a few productive years. If you keep chickens for eggs, you should have a plan for the retired hens: will you keep them as pets or for meat?

○ **EGG.** After a fertilized egg is laid, it takes 21 days for the baby chick to hatch out of the egg.

○ **CHICK.** The chick uses its egg tooth to break through the shell. It grows very quickly in the first weeks. Chicks need special attention because they are vulnerable to illness and predators.

○ **PULLET.** A pullet is an immature female chicken under 1 year of age. Depending on the breed, pullets will usually start laying eggs when they are around 6 to 8 months old.

○ **HEN.** A mature female chicken will lay a certain amount of eggs in her lifetime. The amount varies greatly depending on breed and how the hen is managed. In winter with less natural light (under 13 hours per day), hens will stop laying eggs if artificial light is not provided. During periods of laying eggs, depending on the breed, hens will usually produce an egg every 24 to 36 hours, and as she ages, that time period grows. Hens can be prone to being broody—inclined to sit on a clutch of eggs to hatch them, and acting hostile by fluffing up their feathers and making a grumbling, growling sound as you approach—which is good for people who want to raise chicks, but may be problematic for the chicken keeper not intending to do so. This natural behavior is said to be contagious in the flock and other hens will follow suit, making overall egg productivity drop.

○ **ROOSTER OR COCK.** The mature male chicken over one year is called a rooster or a cock. These lovely birds have impressive plumage. Cockerel is another term for a rooster and most often describes an immature male chicken under one year of age. Roosters are not necessary to keep in the flock, because hens will lay eggs with or without a rooster. If you have a rooster, you will have fertilized eggs and the opportunity to raise chicks. Keeping a rooster has some pros and cons: They are good providers and will guide their hens to food and protect them, but can also become aggressive. While mating, a rooster can injure a hen's back, and it is common for hens to have missing feathers and get scratched up from the rooster's claws. Also, their crowing is especially loud in the early morning, but can also be heard throughout the day, so they are often outlawed in urban areas.

○ **MOLTING.** All chickens naturally shed their feathers on an annual basis starting at about 18 months. Molting typically happens in the fall, lasting for several weeks, and at that time, egg production stops. As the chickens age, the process takes longer.

○ **MENTOR HENS.** Older hens can be beneficial to the flock. They are mentors to younger birds in teaching them how to forage, where to lay, and other behaviors. The older hen will still be productive in the garden as well, producing fertilizer and taking care of pests. She will also lay an occasional egg. To some chicken keepers, keeping older, unproductive hens is not desirable because they require feed and care with no eggs to show for it. Older hens can be kept as pets or culled for stew meat.

For free-ranging chickens, you want to make sure you do not have too many birds on a small piece of land without proper planning and management. Allowing a dozen chickens to free range in a backyard of 1000 square feet could decimate all green plants in a matter of days. But if a garden of that size is well designed for chickens and the number of birds is small—approximately three to five hens—it can work. For gardeners who plan on raising the chickens for eggs and as garden helpers, the chart on page 22 may be helpful in determining how many birds to keep. In general, free-range chickens should have no less than 250 square feet per bird. And I recommend keeping at least three hens, because they are social animals and will not be happy without a small flock of friends.

Selecting Breeds

In choosing what kinds of chickens to keep, it is important to ask yourself why you want chickens. The way you raise them will vary depending on whether you want egg layers or broilers for the table, and whether they are meant to be family pets or the animal you plan on eating after a few short months.

With hundreds of breeds to choose from, you should do plenty of research to find the right breed for your situation. Just like plants, chickens come in different species and varieties. Breeds range in size, temperament, egg or meat production qualities, color, foraging abilities, and more. Some breeds do well in confinement while others do not. Many domesticated chickens originated in specific areas of the world—including Asia, the Mediterranean, Europe, North America—and these breeds have different characteristics because of their original climate and geography, such as heat or cold tolerance.

Not all chickens are created equal for free ranging in a garden setting, and you may have reasons for seeking out a specific breed. I use the term "breed" loosely; there are purebred and hybrid breeds, or strains, available. Hybrids have often been bred to produce well commercially or to behave in certain ways. There are many qualities in chicken breeds that may or may not be desirable for the chickens you keep:

○ *Heritage breeds.* Much like heirloom plants, heritage breeds are a good option for preserving genetic diversity. Many have qualities of self-sufficiency, such as good foraging abilities, resistance to disease, and natural mating habits. Heritage breeds also possess a longer and healthier lifespan than most hybrid breeds. They grow and mature more slowly, which gives them time to develop stronger skeletal and internal systems before carrying all of their body mass.

○ *Egg-laying abilities.* Some breeds are more dependable than others in laying eggs at a consistent rate. Some breeds can lay up to 300 eggs a year, like Leghorns, or as little as 50, like Cornish.

○ *Meat birds.* Breeds known for their quick growth are often processed after a few short months. Because of the quick growth, they can have potential health problems, such as their legs not supporting the weight of their body. In most cases, these chickens are raised for a quick turnaround and are not going to be pets or intended to be free ranged in a garden setting, but they can be kept in other well-managed, confined range systems with access to foraging in the fresh air. Cornish or crosses are the most common types of chickens raised for meat as well as many dual-purpose breeds.

○ *Dual-purpose breeds.* These self-reliant breeds are the most common for backyard chicken farmers because of their versatility, vigorous health, good foraging abilities, and disease resistance. Dual-purpose breeds include Australorp, Brahma, Buckeye, Chantecler, Dominique, Faverolle, Jersey Giant, New Hampshire, Orpington, Plymouth Rock, Rhode Island Red, and Wyandotte.

○ *Temperament.* Chickens can range from being docile and sweet to being flighty and unpredictable, or aggressive, making them difficult to handle. Some chicken breeds can tolerate being confined better than others, while some chickens are better foragers. Some chickens have strong wills while others are more passive. Keep in mind that the chicken's personality may change or be a result of how it was raised. If chicks are handled a lot by humans, they will be more sociable than if they were not.

○ *Color.* Chickens come in a vast assortment of colors and some have beautiful plumage patterns. Many breeds come in a variety of colors to choose from.

OPPOSITE: White breeds are more visible to predators in the garden.

o *Egg color.* Different breeds lay eggs with different colored shells. The most common are white and many shades of brown, but eggs also range from blue to green and chocolate brown. I admit, I keep chickens based on the color of their eggs. Our family loves Dr. Seuss's books, so having real green eggs with ham is a treat.

o *Hardiness and climate tolerance.* Whether you live in Anchorage, Alaska, or Orlando, Florida, you can keep chickens. In different climates, however, you will want to choose different breeds and housing. Some breeds are better adapted to being in colder climates. A good rule of thumb is that the heavier feathered breeds do best in cold. But knowing where the breed originated will give you hints about where it will do well.

o *Ornamental or exhibition breeds.* These breeds are popular for showing and competition. If you haven't been to a chicken show, I highly recommend it. You will see more chicken breeds than you ever imagined, and they are pampered and groomed much like dogs would be for the Westminster Dog Show. Some types of chickens also have unusual features that may have pros and cons for your situation:

o Crested birds, with feathers on the top of their heads, have obstructed vision and may not be as alert to predators. It is possible, though, to give your crested chicken an occasional feather trim to help its vision.

o Five toes are typical in a few breeds, such as Dorking, Faverolle, Houden, Sultan, and Silky. Four toes are normal, and some can even have six.

o Feathered legs and feet. Chickens with feathers down their legs are said to scratch less than their bare-legged cousins. They will certainly still scratch the soil around with their feet in search of food, but how much less? It depends on the chicken.

> ### tip
> *Chickens with a large combs and wattles—the red, fleshy parts on the chicken's head, the comb being on top and the wattle under the chin—are more likely to suffer from frostbite in cold climates. Heavier breeds with denser feathers will do better in the cold.*

OPPOSITE: Licorice the Black Silky hen is a wonderful addition to Angela Davis's garden.

Where to Get Your Chickens

There are many ways to start a flock: you can obtain fertilized eggs, baby chicks, pullets, or mature birds. Depending on the breeds you are interested in, you may need to research where you can get them. Poultry clubs will help you find breeders in your area who sell chickens. You are even likely to find someone selling chickens in classified ads on the Internet. Most people who are new to chickens start by purchasing chicks from local feed stores or from mail-order hatcheries. Before you buy your birds, there are a few things to know.

Chicks can be purchased directly from mail-order hatcheries, where you can pick the exact breeds you are looking for. There is usually a minimum order for purchase, which means you many need to combine orders with a friend if you do not intend to bring home a couple dozen new birds. If they are being shipped long distances, then don't be surprised if you open the box and find a few dead chicks, and be ready to follow instructions carefully.

I prefer to buy and choose my chicks in person at a local feed store, where I can assess their health and make sure they are alert and not the bully of little flock. I watch the store's order-shipment dates and spend a few minutes observing the chicks. When buying chicks, watch for how they are labeled and sexed, for example "straight run" means the chicks weren't sexed and you may end up with all roosters. If they are not labeled, you can assume you are getting all females, but don't be surprised if you end up with a rooster or two.

Rescue and ex-battery hens, or chickens that formerly lived in confinement on a factory farm, are often available. Some rescue organizations take in chickens from owners who can no longer keep them. If you choose this option, consider that the chickens may have not been handled and might be skittish and difficult to deal with. Also, rescued adult birds may harbor parasites and should be confined before introducing them to your flock.

>
> ### tip
> *For free-range birds, it is best to have ones with a dark or camouflage color so they are less easily spotted by predators.*

Large and friendly, the ORPINGTON chicken is known as a good starter hen for backyard homesteaders or families with small children. This attractive breed is a reliable producer of brown eggs (175 to 200 per year). Orpingtons grow to be 7 to 10 pounds, which means they are too heavy to fly, making them a good urban chicken. They go broody easily, making them good mothers.

The RHODE ISLAND RED (RIR) is a utility breed raised for both the table and for egg producing. It lays at least 250 brown eggs per year, and is hardy and disease resistant. These chickens make good pets and are generally very friendly. They can be aggressive toward other chickens if they are annoyed, especially in tight quarters. For urban chickens in a coop and run with no free-range access, the RIRs of the flock may become bullies. They rarely go broody, and are heat tolerant.

PLYMOUTH ROCK chickens, often called Plymouth Barred Rocks because of their coloration, are quite personable and docile. They are prolific egg layers, producing at least 200 brown eggs per year. The Plymouth Rock breed comes in many colors. Because of their heavy build, they don't tend to fly over fences. This breed tolerates cold weather well.

Bantams are often called the miniature chicken breed, because they are typically one-fourth to one-third the size of their full-sized counterparts. They are known for being quieter than full-sized birds, and for just about every full-sized breed, there is a bantam version of that breed. There are several true BANTAM breeds, including SILKY, JAPANESE BANTAM, PEKIN, and SEBRIGHT, which have no large-breed counterpart. Bantams can be flighty and are known to fly over tall fences, so consider clipping their wings or have proper fencing for containing them. Their egg production is not always as good as that of larger birds, and they have smaller eggs.

The SILKY, a bantam-sized ornamental breed, has five toes and is known for its unusual hairlike feathers. Silkies come in many different colors and have calm and friendly personalities, making them good pets for children. Their eggs (at least 100 per year) are small, so you'd need two eggs for recipes calling for one large egg. They often go broody, and make excellent mothers. Their skin and bones contain a blue-black pigment, which in Eastern medicine is believed to have healing properties.

People often confuse the three breeds AMERICANA (shown), ARAUCANA, and EASTER EGGER, because they all lay bluish-green eggs (about 200 per year), but they have distinct differences. The Americanas and Easter Eggers are great foragers but don't hesitate to fly over fences, so keep that in mind when you are planning your garden fencing, or otherwise clip their feathers. Americanas are distinctive in their tufted cheek feathers, pea comb, white skin, and slate blue legs. Araucana chickens are rare and have distinct ear tufts, no rump feathers, green legs, yellow skin, and they lay blue eggs. Easter Eggers (EE) are considered a "mutt" breed and can come in any color and lay any color of egg. Any chicken with the blue egg gene is considered an EE, which is classified as nonstandard by poultry associations.

The hybrid "sex link" chickens are known for their outstanding egg laying ability (at least 300 per year) and the fact that their color indicates their sex. There are different breeds of sex links, which can be easily identified by sex as chicks by feather color, and are called by other names depending on the color and breed: BLACK STAR (shown), RED STAR, CALIFORNIA WHITE. They are cold hardy and have an easygoing temperament, making them a good choice for backyard homesteaders.

The OHIO BUCKEYE breed has a reputation for hunting mice that rivals that of cats. They are very friendly and are great foragers, making them an excellent free-range chicken.

tip

Good foraging breeds include Ancona, Dominique, Sussex, and Wyandotte, as well as other favorites, Ohio Buckeye, Orpington, and Rhode Island Red.

Bringing Home the "Biddy Babies"

Baby chicks are irresistibly adorable. Have a brooder box (a chick housing separate from the older birds) set up and ready, so the "biddy babies," or the hen's chicks, have everything they need for a good start. Chicks are susceptible to illness more than larger birds, and are sensitive to cold and damp conditions, so keep them in an area that is draft free and out of the elements. I keep chicks in the garage for the first several weeks, where I can easily watch and take care of them. My brooder box is a large wire dog crate that we place wood panels up against and adjust as the chicks get larger and need less warmth and more ventilation. Baby chickens need a few basic elements to be healthy:

o *Safe housing.* A brooder box is a simple structure and can be as simple as a cardboard box. Chicks grow fast and make a mess, so be ready for that as they grow. The brooder box needs to be safe from predators, possibly even your dog or your cat.

o *Warmth.* For the first week of a chick's life, the ideal temperature is 95ºF, and then every week, lower the temperature by 5 degrees. If you use a heat lamp, be sure that you can adjust its height so the chicks won't be too hot or too cold. Chicks are good at showing you how they feel about the temperature: if they are too hot, they'll move away from the lamp, and if they are too cold, they huddle underneath the lamp.

o *Fresh food, air, and water.* Chicks need fresh food, air, and water at all times. The chicks will likely make a mess of the food and water and will need to be cleaned up after. There are special formulations of food for chicks, many of which are medicated to prevent illness. Bedding needs to be changed regularly, especially when it becomes saturated. Good bedding choices for chicks are white wood shavings or shredded paper.

Just Like Plants, Chickens Need Time to Harden Off

As the chicks grow, they will need to be introduced to the outdoors slowly and at the right temperatures. I start chicks in the spring, so by the time they need to be acclimated to the outdoor temperatures, they will be kept warm naturally. I start by taking the chicks' entire wire dog crate and placing it in a grassy area without the bottom when it is at least in the 50s or 60s and preferably calm and not windy or rainy. I usually do this on a weekend when I have some gardening to do and can keep a close eye on them. The chicks will learn how to search for insects and eat the grass.

You can use temporary fencing to enclose the chicks as they get older, but be sure they are safe from predators and have food and water. Once the chicks have their feathers, they can be introduced to their coop or night house. They should be locked in for a couple of days to learn that this is their new home. Then, take time to acclimate them to the garden, so they get into a routine that works well for them and you.

Feeding Your Chickens

What, when, and how much to feed your chickens will depend on several factors: the age and type of bird, whether it is kept in confinement or allowed to free range, and the seasonal changes in weather. If chickens are confined, they will completely depend on you to feed them, and birds allowed to free range will be able to forage for some of their food on their own. Chickens have different nutritional needs as they develop and if they are being raised for eggs or for meat.

o *Feed.* Commercially available organic feeds and supplements can be purchased from local feed stores. The feeds have different formulations and are designed to fit the needs of birds of different ages. Many feeds are sold as a complete food containing everything a chicken needs, but the hens may require additional supplements. Be sure to read the labels.

o *Calcium.* Laying hens need additional calcium once they start producing eggs, which can come in the form of oyster shell, crushed bone, or you can use cleaned and crushed egg shells from their own eggs.

o *Grit.* All chickens need grit to help them digest their food, because they have no teeth to grind the food. Grit can be found naturally as they forage in the soil, or can be purchased as bagged granite dust or other crushed rock material.

o *Greens.* Greens are another important part of a chickens diet. These can come from grass and other plants they forage, or in the form of kitchen scraps or weeds from the garden.

o *Snacks.* Many people dish out treats for their chickens—scratch feed is a common one. Just be careful not to feed them too much; moderation is key. Other healthy treats to feed your chickens include plain yogurt; the probiotics are good for your chicken's digestive system, and they love it. Chickens will also devour insects you've collected from your garden. When I garden, I keep a pail nearby and collect pests such as snails, slugs, and cutworms, and offer them to my chickens to gobble up.

I prefer to feed my chickens a handful of scratch in the morning as I let them out to range, then an organic feed once a day when they go to bed. I prefer not to "free feed" (leave food in their feeders at all times for them to eat as often as they want), since this approach can invite rodents to dinner and obesity in the hens. Everyone's situation and feeding routines will vary, just like chicken keeping and gardening.

an important note about cedar and pine shavings

While cedar and pine lumber are considered great antifungal, insect-repellant materials, the shavings of these woods release volatile compounds (for example, aromatic hydrocarbons, phenols) from their oils that are not healthy for a chicken's sensitive respiratory system. These emissions can be especially dangerous to chicks, and to chickens that are confined for long periods with minimal ventilation and limited access to fresh air. Cedar and pine plants or objects made of cedar and pine lumber, however, pose little threat to chickens.

incorporating new birds into an established flock

Chickens can be downright feisty and mean toward each other, especially toward newcomers, who can get picked on to the point of physical harm or even death. It is important to introduce new chickens carefully to a flock to minimize problems from happening:

1. Start by putting the new chickens in an area that is still visible to the other birds but keeps them physically separated, such as a chicken tractor or a dog kennel. Let the new and old birds see each other for a few days, which is also a great time to make sure the new birds aren't bringing any parasites or illness to the flock and lets the new birds learn the feeding and cleaning routine.

2. After a few days of physical separation, introduce the new birds to the flock in the evening when the flock is settled in and roosting and everyone has been fed separately. This is when resident birds are least active and less likely to harass new birds.

3. Stay and watch the birds interacting during the first 15 to 30 minutes, so you can see if a bird starts to bully the new birds. That way, you can step in to separate the birds if you need to, although they will probably quickly settle their differences themselves.

I've used this system successfully with many introductions of young pullets and orphaned adult birds into an established flock. Make sure there are enough feeders and waterers for all, and the new birds have space to get away from a bully if needed.

BEING A GOOD NEIGHBOR

I think it is safe to say that we have all had an unpleasant experience with an obnoxious neighbor. Whether it is a loud party or a barking dog, when your environment has been tainted by someone else's unwanted noise, it is easy to get frustrated. Noise isn't the only thing we need to keep in mind with chickens. Some birds are quieter than others, but all of these lovely little ladies poop—a lot—and their coop needs to be cleaned on a regular basis or the air can become rancid, especially if they are shut in the coop all day every day. While free ranging, they can also go onto other people's property, and while we may enjoy their antics, other people might not. Fencing is critical in making sure your chickens don't become someone else's dinner. We love our hens, but we should never assume that our neighbor shares our enthusiasm. Don't let chicken keeping get a bad name because of irresponsible ownership. Considerations include:

○ *Noise.* Before you get your chickens, stop and listen to what you hear in your backyard. Is it chirping birds and leaves rustling in the wind? Or is it the loud hum of traffic or barking dogs? If you detect background noise, adding a few chatty hens will hardly make a difference. Most hens make some noise, especially right after laying an egg, but nowhere near the decibel level or consistency of a rooster's crow. Throughout the day, hens softly cackle, which sounds like a mumble and varies from bird to bird and breed to breed. Some breeds are known to be quieter than others.

○ *Smell.* A well-maintained coop will have little if any smell. The chickens themselves don't smell one bit. This might be the first concern you might hear from neighbors even before chickens arrive in your garden, so it is important to not let this type of complaint become a reality. Having a regular routine and compost system for the birds' manure in place is crucial. If the birds free range, then their manure will be spread out over a large area and will dissolve quickly into the soil.

○ *Attracting disease or pests.* Without chickens in our backyards, we are all susceptible to pests or disease depending on our lifestyle and the health and management of our household. Chickens are less likely than we to get sick or catch a disease transmittable to humans. Common chicken illnesses are preventable by good management, and many hatcheries vaccinate chicks against illnesses before they are sold. Without a good setup and proper maintenance, pests could become a problem, so be aggressive in preventing and controlling any issues when designing and monitoring your chicken system.

The bottom line is that proper management is key in maintaining a healthy, neighborly flock.

is it legal to keep chickens?

Local land use codes determine whether you can keep chickens in your jurisdiction or community. Many major cities have laws stating whether chickens are permitted or what kind of restrictions are in place, such as quantity of birds or coop placement parameters. If you live in a neighborhood where you pay homeowner dues to a homeowners association, then you are likely to have a set of Covenants, Conditions and Restrictions (CCRs) that list specific rules. To find this information, start with a search of local jurisdiction websites on the Internet.

If you cannot find the information you are seeking, call your town or county planner.

If you live in an area where chickens are prohibited, then a community group is probably actively trying to change the law. The Internet is a good place to start, or talk to the staff at a local feed store to find out what the buzz—or clucking—is.

OPPOSITE: Chickens forage in an ornamental garden.

SEASONAL CONSIDERATIONS

At different times of the year, it is important to be aware what your chickens face being outdoors and how your garden may be affected by the chickens.

o *Spring.* Spring is a great time to purchase chicks and to apply compost that has been curing over the winter. New plants are tender and can be easily damaged by chickens scratching or pecking. Be careful to protect newly sown seeds and emerging perennials that can be easily damaged. A curious chicken looking for a tasty morsel can knock off a tender emerging shoot and destroy the plant's new growth. Use temporary fencing or other physical barriers for vulnerable areas.

o *Summer.* The heat of the summer means we need to make sure our chickens are cool, have shady shelter, and have fresh water at all times. In the coop, air circulation is important. In the garden, most herbaceous perennials should be up and out of harm's way. Ripening crops may need protection during their peak production, but possibly not all summer long. I find this time of year the best for doing a thorough coop cleaning since the chickens can be outside for longer periods and the warmth speeds drying time.

o *Fall.* Autumn is a perfect season to put the chickens to work. Instead of tediously weeding and cleaning up beds after the growing season, let the chickens do the work. If it is only in specific areas, use a chicken tractor. Fencing can come down in areas that were being protected as crops ripened, and compost from the spring can be used as mulch to protect plants and soil from winter rain and cold.

o *Winter.* Many chicken owners allow their chickens full access to the garden all winter long. In this season, there are fewer delicate plants in the garden, and there are insects to eat and soil to scratch in. Because of freezing temperatures, make sure the chickens always have access to fresh water and housing to protect them from the cold.

OPPOSITE: This chicken is not so sure about the snow.

CHICKENS IN YOUR GARDEN

practical considerations

Deciding to bring chickens into your life and your garden will no doubt bring about some changes in your lifestyle. There are many decisions to make, from choosing which kinds of chickens to get to designing their housing, and you will need to create a system for managing them in your backyard. You must decide which of their attributes you want take advantage of, and what you will do to protect the birds as well as how to protect your plants. With some planning, you can get your chickens and your garden into a smooth coexistence.

the
zumwalt chicken ranch

MEET ELIZABETH ZUMWALT, 9 years old, chicken whisperer prodigy, chicken educator, and entrepreneur. She writes a blog about the family's Bantam hens and her business selling chicken eggs to clients. She also takes the hens in a chicken tractor to public parks and local destinations to educate people about chickens. She gives half the money she raises to charities of her choice every month. Her mother, Missy, active at the local elementary school, is the garden club coordinator and is working on making chickens a part of the regular education curriculum.

The Zumwalt family lives on a small urban lot where the neighbors are fairly close on the other side of the fence. They chose Bantam breeds because of their reputation for being quieter than their larger cousins. The chickens free range all day in the backyard. In an effort to keep neighbors from complaining, they positioned the coop close to their master bedroom window, so if there is any noise they'll be the first to hear it. The family raises chickens instead of conventional pets like dogs, cats, hamsters, or fish, which can require more care and the kids can eventually lose interest in. The Bantam hens have names and even get to play dress up on occasion.

They have renovated their once-overgrown backyard landscape to create a more family friendly space with a beautiful, chicken-friendly garden. The patio was extended, and the new plants included many ornamental species with a good mix of evergreen and deciduous types, plus layers of perennials and groundcovers. Designated areas of edible growing space are close to the kitchen but not fenced off from the chickens. The chickens get kitchen scraps, except for food like tomatoes or blueberries that they wouldn't want the chickens to eat straight off their plants.

Nearby predators include a family of bald eagles that nest a few blocks away, but the excellent coverage in their garden protects the chickens. During their garden renovation, many of the large, overgrown shrubs where removed, leaving lots of bare ground and no place for the chickens to take cover. One day Missy witnessed a large bald eagle land in the backyard and fly away with one of their Bantam hens. Since the landscape was completed, they haven't had any incidents with predators, but nearby cats are a concern. Fortunately, Bantams are quick and agile and can fly to get away if they need to.

The chickens have been a wonderful addition to the Zumwalts' garden, and the birds have taken chores like weeding off the family's plate of responsibilities. There are some plants that need protection in the spring because they come up early and slowly, such as hostas and some grasses. When the chickens scratch mulch into pathways, Kurt says, "It's no big deal. You just sweep it back."

OPPOSITE: The Zumwalts' garden has a good mixture of plants for a chicken garden: edibles, ornamentals, evergreens, and deciduous plants. NEXT PAGES, CLOCKWISE: Elizabeth pulls the mobile chicken trailer used to transport the hens for educational events. Elizabeth Zumwalt loves her hen Lucy. The chickens graze in their mini garden. The Zumwalt family sits around the patio table (left to right): Jon (7), Kurt, Missy, Jimmy (5), and Elizabeth (9), with their chickens nearby.

BENEFITS OF FREE-RANGE CHICKENS

Allowing your chickens to free range at times will undoubtedly lead to happier and healthier birds compared to chickens confined to enclosed quarters. The fresh air, sunlight, and access to fresh sources of protein and greens will give the chickens a life closest to their natural habitat. Because chickens originated in Southeast Asia and India and traditionally foraged on the forest floor in the tropics, they can be self-sufficient creatures. We can let them behave as they would in the wild by letting them out of their coop, but they need to be managed.

Before you just open the coop door and let the birds out, you will need to look at your garden like an ecosystem. My garden design philosophy has always been to create gardens that mimic Mother Nature as closely as possible. My method recognizes the balance of organisms both in the soil and above ground. In every healthy ecosystem, there is a predator for every pest and interconnected relationships among species of plants, animals, and organisms that create balance.

This web of life doesn't require human intervention, such as the application of pesticides and fertilizers. In fact, applying such chemicals throws the ecosystem out of balance. In many gardens, we try to attract wildlife and specifically birds to help control insect populations. By keeping chickens, like the wild birds we want to attract, they will help keep our insect and pest population in check. Chickens can benefit our gardens in many other ways, too, as long as we don't have too many birds for our garden ecosystem. We need to find a balance.

managing your chicken manure

While chicken manure can be a great benefit to the garden if managed well, it can also be a problem. If left to pile up in coops and runs, chicken manure can attract flies. It can also produce excessive ammonia, which is not healthy for chickens to breathe and creates a stench that will give reason for your neighbors to complain. If manure is left in large accumulations in a raw state, it can also pollute the soil and water from rain runoff. Make sure that your chickens' manure is cleaned up regularly and composted correctly.

A Food Source

What attracts most people to the idea of keeping chickens is gathering fresh, delicious eggs from their backyard coop. Reconnecting with where our food comes from is becoming more important in people's lives, and chickens give families a way to teach their children about that connection. Chickens that free range in organic gardens are believed to produce healthier, more nutritious eggs. Mother Earth News completed a study in 2007 that compared eggs from truly free-range birds to USDA nutritional data, and found that the free-range eggs may contain:

- 1/3 less cholesterol
- 1/4 less saturated fat
- 2/3 more vitamin A
- 2 times more omega-3 fatty acid
- 3 times more vitamin E
- 7 times more beta-carotene

While the factory farming egg industry disputes these findings, there are many studies with similar findings that date back to the 1970s.

Not all chicken owners raise birds for meat, but it is always an option. Some sources of information on this are in the Resources at the back of the book. Homesteaders who raise and breed chickens can have a steady supply of meat and eggs, if they manage the flock well.

too many eggs?

Sometimes we chicken keepers have more eggs than we know what to do with. You can give them to neighbors and coworkers, sell them, or barter them for other goods. Here are more ideas:

○ GIFT THEM. A dozen of your freshest eggs make a nice hostess gift. Don't forget to personalize the egg carton with your own label or even ribbons.

○ PRESERVE THEM. Did you know that you can freeze eggs? You separate the whites from the yolks or blend them together, and put them in a freezer-safe container to use later. But don't freeze them in the shell. You can also pickle eggs.

○ GET CREATIVE IN THE KITCHEN. If you're tired of the same old egg recipes, browse the Internet for new ideas and inspiration, and then create your own egg cookbook.

OPPOSITE: A Buff Orpington helps out in the compost area.

Manure and Compost

Manures have been used for centuries as a natural source of nutrients for soil. Free-range chickens deposit their manure directly onto the soil, where it is quickly dispersed, reducing the time we spend cleaning the coop. Organic matter is an important part of healthy soil, which is the foundation of organic gardening; it creates good soil structure and provides nutrients for microorganisms. Chicken manure is one of the most nutrient-rich livestock manures available to gardeners. When you keep chickens in a permanent coop and run, the manure and bedding can be easily gathered and composted to create a rich soil amendment and mulch for our plants.

Composting involves a natural process of organisms breaking down material just as they would in the top layers of soil in nature. If we provide the organisms with what they need to do their job, we can have a rich finished product that gardeners would pay a lot of money for. Composting can be a fairly simple process with low labor input, or a highly involved endeavor, depending on how quickly you want results and on the volume and quality of raw materials you are using. There are many approaches to successful composting. Most common residential systems will have a couple of bins (each 1 cubic yard) where the material can be rotated during different stages of composting. As one pile is getting started, another is finished, or cured. There are several things to keep in mind.

○ *Space.* How much space will you need for composting? This depends on how much waste your garden generates and how many chickens (or other animals with manure and bedding) you keep. A typical quarter-acre lot can easily keep two compost bins, each 3 feet square, working on a year-round basis.

○ *Weather and moisture.* Composting material should be kept moist but not saturated. In some climates, it may need to be watered or covered, depending on the weather.

○ *Heat.* For "hot composting," a good internal temperature for your compost is between 130ºF and 160ºF degrees. As the organisms in the compost pile break down the material, heat is generated, and the temperature can be used as an indicator of how well your compost pile is working. The heat helps kill pathogens and weed seeds, but if the compost pile is too hot, beneficial microorganisms can die. Once the pile has been hot for several days, use a pitchfork to move the hot internal material to the perimeter of the pile and the cooler material to the center. A long thermometer can be used to check the internal temperature of the compost pile. A "passive or cold" compost method does not rely on heat and takes less labor and management, but the process takes longer and does not always kill the pathogens and weed seeds.

○ *Air and layering.* The composting organisms need air to breathe and work just like we do. It is best to put the different materials down in layers to maximize the airflow, allowing the organisms to work efficiently. You can do this by turning the pile with a pitchfork or you can let your chickens scratch and peck at the material to distribute it. A perforated pipe placed in the middle of the pile is another way to provide more air circulation.

○ *Particle size.* We can compost material of different sizes and at different states of decomposition. In general, the smaller the particle size the quicker it will decompose. Herbaceous material will decompose faster than woody material, so it is best to shred or chop large, woody material before adding it to a compost pile.

○ *Rodents.* Rotting vegetative materials may attract pests such as rats. Putting food scraps in an open composting system can create a rat habitat that can quickly get out of hand. Food scraps can be given directly to the chickens instead or placed in closed systems such as worm bins.

Many types of manufactured composting bins are readily available for purchase. They have different features: some tumbler varieties allow easy turning of the material inside, some are designed to be a specific shape, size, and color. But why not make your own? There are easy ways to create a compost pile, for example, by simply piling your compost into a neat 3 by 3 foot square and layering the organic material. Other easy ways to create inexpensive compost bins, which are easy to access and can be reused over and over, include:

○ *Pallets.* Take four wood pallets and stand them on end, making a square. Attach them together with hardware similar to a gate hook and latch or bungee cords, so the pallets can be taken apart easily.

○ *Metal fencing.* Create a 3- to 4-foot-diameter circle of metal fencing, and tie the ends together, making a round.

compost ingredients

The basic compost ingredients are materials that contain carbon and nitrogen. The ideal ratio is somewhere around 30:1 (C:N). The ingredients are brown materials (carbon) such as bedding versus green materials (nitrogen) such as manure.

CARBON (brown, dry material)	NITROGEN (colorful, wet material)
Sawdust	Manure
Straw	Veggie scraps
Leaves	Coffee grounds
Pine needles	Alfalfa
Wood chips	Garden clippings
Newspaper	Grass clippings
Corn stalks, peanut shells	Weeds
Cardboard	Alfalfa, hay

tip

When your compost is ready for use, be sure to save a small amount as a startup innoculant to get your next pile working quickly.

more about chicken manure

Without being composted, chicken manure is too hot for plants, especially young, tender plants that can easily burn and then die. Chicken manure should be aged or composted for 60 to 90 days before using it in the garden in large amounts.

○ **MANURE TEAS OR LIQUID FEED.** While many gardeners swear by manure teas or liquid fertilizer, homemade brews potentially can harbor harmful bacteria and pathogens if they are not prepared properly. It is best to do your research and use composted manure and an aerated tea system if possible.

○ **A SIMPLE LIQUID FERTILIZER RECIPE.** Soak 1 part composted chicken manure in 10 parts water in a large lidded container. The manure can be placed in a burlap bag or old pantyhose and weighed down in the container with a rock. Let soak in warm weather for up to 1 week, and discard the soaked manure on the compost pile. Before using it in the garden, dilute the liquid again at 1:10, and apply to the soil.

BELOW: Pallets (left). Metal fencing (right).

Pest Control

Chickens need protein, and thrive on a diet including all sorts of insects, whether crawling or flying, and they especially love insect eggs. Their taste for creeping and crawling critters has been known to wipe out devastating populations of insects such as termites, grasshoppers, and ticks. In gardens, chickens will patrol for anything moving around in the soil, including grubs and worms. They will clean up any fallen fruit that harbors and can overwinter pesky insects that can cause a problem the following year. One season, I fenced off an area of plants that my chickens would otherwise free range over, and that year, I ended up having more pest problems in that area than I had in the past. Chickens do a wonderful job of keeping pests in check.

Garden Helper

Chickens can be helpful to the gardener by tackling some of the more tedious and backbreaking work.

o *Weed control.* A newly emerged weed seedling doesn't have a chance against the continuous scratch of a chicken determined to find a juicy morsel in the soil. Either the scratching will rip the weed out of the soil, or the chicken will nibble at the weed. At certain times of year when weed growth is hard to control, you can simply let your chickens do the work for you. For unprepared gardeners, this practice can get chickens in big trouble, because chickens can't differentiate weeds from the plants we adore and the seeds we've planted. Chickens will usually leave mature plantings alone, and will weed around them and in pathways or other areas without established vegetation.

o *Aeration.* Chickens scratch in the soil for food, turning it over and keeping it aerated, making the top few inches loose. This behavior is helpful in an area with a good layer of mulch, but in compacted mineral soils, chickens may tend to create holes with their scratching and dust bathing. Also, chickens can help you turn and aerate your compost pile by spreading it out, and depending on your system, you may just need to re-pile it when they are done.

o *Lawn removal.* Lawn removal can be a daunting task. Renting a sod-cutter, digging it out with a shovel, and sheet mulching can require lots of physical labor. So why not let your chickens do the work for you? With a chicken tractor, the girls can spend several days or weeks scratching at the grass in one area. With this focused attention, they can decimate what is left of any lawn. When you place chickens in a tractor system to remove lawn in an area, be sure to add a layer of bedding (wood shavings, chips, leaves) every few days, to keep the soil from becoming compacted and so the soil can be enriched with the chickens' manure and aerated by the chickens' scratching, leaving the space ready for planting after one or two seasons.

o *Lawn mowing.* Chickens graze best when your lawn is maintained at a height between 2 and 5 inches. If the lawn is taller, they can't reach it to graze, and below 2 inches, they can damage the lawn. An added bonus, they fertilize the lawn while they are "mowing" it.

Chickens as Composters

Chickens do a wonderful job of eating just about anything we give them and quickly turning it into manure. Some municipalities are finding that chickens can lessen the amount of solid waste a community creates and are encouraging households to use the birds as feathered garbage disposals. This step could decrease taxes and lower the carbon footprint of cities. But we shouldn't give them every scrap of food and yard waste we have.

---

yum, slugs!

Slugs are a particularly frustrating for many gardeners. Most chickens love to eat slug eggs, but they normally don't eat slugs whole. It is possible, however, to teach them to love the taste of slugs, and eventually make them into slug-eating machines. Here's how:

o Collect slugs in an old container.
o Using old scissors, chop them up into small pieces.
o Feed the chickens the smallest pieces first, and work up to larger pieces.

The chickens will eventually love to chomp down an entire slug.

OPPOSITE: You can use a wedge feeder or a small suet cage to hold the chicken's food in place.

○ *No-no's.* Keep in mind a chicken's natural diet: they don't naturally eat processed foods like bread, so give those sparingly. Only feed chicken-sized bites of food or they may have trouble eating it or get an impacted crop; the crop is a storage pocket at the base of the chicken's neck where it initially stores food before it moves through the digestive system. Yard waste from our weeds and trimmings can be nutritious, but do not feed them poisonous parts of plants. And for free-ranging birds, you may find that if you feed them left-overs of your favorite vegetable or fruit in the garden, such as berries or tomatoes, they may seek it out and take it right off your plant, so be careful not to give them snacks you don't want them to eat in the garden.

Chickens should not be fed or given anything with mold, alcohol, caffeine, or foods high in salt. Old feed can easily get moldy and is often a cause of illness and death in chickens, so don't buy more than you can use up in a few weeks. Chocolate, pits from stone fruits, and uncooked or green potato parts are also no-no's. Apple seeds in large quantities and tomato and avocado leaves and stems are also toxic. Onions and garlic in large quantities will make their egg yolks taste weird.

 KEEPING YOUR PLANTS SAFE

Chickens can destroy plants in many ways. They can eat foliage, and they can scratch at the soil and up-root plants in search of tasty insects in the soil. While scratching, they can also bury plants or erode the soil around the base of a plant, exposing the plant's sensitive root zone. We can time our garden routines and plan for chickens to be in or out of specific areas at certain times of the year when plants are most vulnerable. And we can also protect individual plants by using barrier systems or deterrents.

When Plants are Most Vulnerable

Plants are more easily damaged by chickens at various stages of growth, different times of year, or because of their placement in the garden:

○ *New plantings.* Chickens love freshly dug soil. As if they can smell the unearthed organisms in the soil after you've planted, they head straight for recently disturbed dirt. So any new plant is a target for aggressive chicken scratching. Because these newly placed plants haven't rooted yet, they can be easily pulled up or knocked over when a chicken tries a taste of the leaves. Keep an eye on newer plantings, and if a chicken pulls back the soil and mulch, put it back in place. Even if you have to do this a few times, the chicken will eventually move on and the plant will become established.

○ *Young seedlings.* Starting seeds in the ground without protection is a brave act with free-range chickens. They love eating seeds that are easy to find, like beans and corn. When you have chickens, it is wise to do as much of your seed starts indoors and plant them out-doors when they have some growth, or plan to use a barrier method to prevent the chickens from eating the seeds while they are germinating and getting estab-lished. New lawn seed will also need protection.

○ *Early spring.* After the long winter when chickens have foraged every last bit of green, any newly emerging green growth will look like chicken candy. Even if the birds wouldn't normally eat the plant, they will certainly take a little nibble to test it out. For plants like hostas, this attention can leave the bold leaves full of holes. Al-though hail can do the same thing, you can prevent your chickens from damaging leaves with a simple barrier.

○ *Established plants.* Chickens allowed to free range generally will not try to eat large, more established plants. But they will scratch in search of tasty morsels in the soil at the base of the plant, and may leave a hole where sensitive roots could suffer drought stress. Watch for this, and know your chickens' habits during the summertime.

○ *Hillside plantings.* Chickens will scratch and forage in sloped or steep areas, so plantings in these areas can get buried if a chicken chooses to scratch uphill of a plant. Not all plants will mind a bit of extra soil, and in fact some appreciate it. But for many plants, it can cause rot, decline, or can kill the plant. The base of woody shrubs on a slope should be checked to see whether chickens have altered the mulch levels. I rec-ommend planting a dense, spreading groundcover in sloped areas where chickens forage.

OPPOSITE, TOP TO BOTTOM: An old trellis can be used as a ground barrier to protect seedlings. Wire fencing can be used over seedlings for protection.

Barrier Methods and Other Deterrents

The simplest way to protect particular plants is to create barriers so the chickens can't get at them. Gardeners can use various kinds of barriers:

○ *Fencing.* Fencing comes in many types and can be used to prevent plants from getting damaged. Either with a perimeter fence or a temporary, localized fence, large spaces or single plants can be protected. After sowing seeds, I often put fencing down on the ground over the seeds. I use wire mesh and simply bend the edges under to prevent chickens from scratching in that spot. Sunlight and water can still get through, and when the seed germinates and starts to grow, I bend the wire fencing so it's higher.

○ *Stones.* Placing stones around the base of plantings will prevent chickens from scratching. For herbaceous perennials, it is worth spending a few minutes in early spring to carefully place some cobbles in a ring around the plants. Later in the summer, stones can be placed at the base of plants, which will absorb the heat from the sun all day and transmit it to the soil at night.

○ *Sticks.* To prevent chickens from nibbling on tender plants, keep a few branches handy and build a small teepee over the plant growing in the spring. This works well for emerging delicate herbaceous perennials.

○ *Cloches.* On double duty in a chicken garden, cloches protect individual plants from the chickens while offering the plants shelter and warmth from cold environments. Classic cloches are made of glass, but can be made from a variety of materials like sheet plastic or can be as large as movable cold frames.

○ *Netting.* Bird netting used to protect fruiting shrubs and trees from wild birds when the harvest is ripe can also be used for chickens. This practice is common among fruit farmers because birds and other wildlife can detect ripe fruit such as blueberries or strawberries and will come for the pickings.

○ *Water.* Chickens don't like being sprayed with water. A sprinkler or spray from a hose is an acceptable, harmless way to get chickens out of a particular space, when you happen to be there. When you are not around, motion-sensing sprinklers can be effective in preventing damage from foraging animals such as chickens and even deer.

○ *Odor repellants.* While odor as a deterrent may discourage some animals, I have yet to see it work well for chickens. Most odor-based repellants wash off after a few rain events. There are currently bird-proof sprays to prevent wild geese and ducks from entering areas with ponds and large public lawns. The product is sprayed onto the grass, the birds eating the grass get sick from it, then they learn that the grass from that area tastes bad and they move on to another site. But these products should not be used for chickens.

○ *Containers.* Garden containers can be a good way to get plants up and out of the reach of chickens.

○ *Trellises.* With plants trained to grow up and overhead out of reach, free-ranging chickens will find shelter below, while happily foraging and fertilizing the soil.

KEEPING YOUR CHICKENS SAFE

When keeping any domesticated animal, we should be aware of any potential safety risks because we are now responsible for them. Chickens are an easy catch for many predators, so we need to protect them from a potential attack. Proper fencing is the most important thing we can do to ensure our chickens' safety, but we need to acclimate them to being outside and on their best behavior, which will come naturally in the appropriate environment. We can also train chickens to be easy to catch and work with, making them more manageable in our gardens.

Fencing is a critical part of gardening with chickens. Fencing can be a beautiful backdrop to any landscape and can serve more than one purpose, making it an important garden design element. It can also be an eyesore or it can be unreliable, resulting in the loss of your birds or damage to your garden. Fencing can direct our movement and circulation in a garden, moving us comfortably from one area to the next. It can also alter the way the wind and sun flow through your landscape, which can affect your plants and living spaces.

In all living situations with chickens, free range or not, you should have predator-proof perimeter fencing for the birds' protection. Creating a stable barrier between your chickens and the outside world is imperative. It has the added benefit of keeping them in the yard so they do not run off into the neighborhood, attracting predators

OPPOSITE, TOP LEFT: A motion-sensor sprinkler can be used to discourage chickens from getting near plants. OPPOSITE, TOP RIGHT: You can use temporary fencing around the perimeter of your vegetable garden when maturing crops are most vulnerable. OPPOSITE, BOTTOM: An electrical netting fence powered by a solar transformer makes a good barrier.

or nibbling on neighbors' plants. Predators are everywhere, and the risk of having a neighborhood dog or raccoon around is fairly high. Many determined predators will try hard to get at chickens if given the chance.

There will be areas in your yard that you may want to section off from the chickens' access. If your delicate edible plants and vegetable garden are in a constant state of rotation and production, you should really consider fencing the perimeter of that space. If there is a children's play area and you do not want little feet stepping in droppings, then fencing for that area should be considered.

There are many types of fencing available, in design and materials. Do some research, because this item can be one of the more costly elements in the chicken garden. Some factors to think about while making your selection:

o *Function:* Safety is a priority. Is the fence to be multipurpose or not?

o *Specifications:* Write down the height, length of perimeter, number of gates, and so on.

o *Style:* Decide on materials, type of posts, paneling material, colors.

Before you settle on fencing design and materials, ask yourself the following questions:

o *What level of security do you need for keeping predators away from your chickens?* Before designing predator-proof fencing for your garden, you need to identify the types of predators in your area. Never underestimate the animal that may go after your chickens: they are a lot stronger than we give them credit for. To keep ground predators at bay, such as dogs and coyotes, you could pour a submerged barrier of concrete under the perimeter fencing to keep them from digging under the fence. While that may be expensive, you will sleep better at night. Another more common, less costly option is to bury a bottom portion of the fencing so a predator cannot burrow underneath; typically 1 foot of fencing underground is preferred, burying it on the outside of the fence. Concrete pavers can also be placed along the perimeter of the fenced area, but a large dog can easily dig under a paver or push it aside to get at chickens. Large pavers can be effective if placed under a heavy structure like the coop itself. The longevity of the fence material should be taken into consideration, because over time it will deteriorate in contact with soil and water. Another option is to have a hot wire, or an electric fence, at the base of the fencing so predators will get zapped if they try to burrow. An electric wire at the top of the fence will help prevent predators from climbing over to get to the birds. For flying predators such as raptors, overhead fencing may be necessary, especially if you don't have shelter for the chickens to escape to.

o *How high does the fencing need to be?* A typical wooden privacy fence is 6 feet high, while wire mesh fencing materials come in a range of 2 to 5 feet high. Most chicken fences are effective at 4 feet high. But remember, the smaller the bird, the higher it can fly, and *there are always exceptions.* If you do not plan to clip your chickens' wings, a 4-foot fence may not keep them in or out of an area, depending on the breed. While most dog breeds can't jump over a 6-foot fence, I had one that could scale an 8-foot fence with no problem. For other wild animals that can be a problem in your garden, such as deer, you may need a fence of a specific height.

o *Is the fencing multipurpose?* A fence can serve many purposes. Not only will it keep your chickens and garden safe, but it can provide privacy from neighbors, it can be a great barrier to block unpleasant views or sounds, and it can create a vertical space to grow plants, like vines or espaliered trees. If there are other pets being kept in the same area, like livestock, they may require a different type of fencing. The larger the animal, the stronger the fencing material must be, and the openings in the fence can be larger.

o *Do you have a preferred style of fencing?* Fencing can be a major part of the garden's structure and design. It can be customized to your taste and budget. There are many styles to consider, from formal to eclectic. If you like a fencing style that isn't chicken-safe, for example, a post-and-rail pasture fence, you can add wire mesh to it. Most standard fencing is 6 feet tall, making it the perfect height for privacy. But fencing with solid paneling may limit the light and airflow through the enclosed space, which could affect your plants. A full privacy fence can also limit a chicken's curiosity, so they won't want to get to the other side. Wire mesh fencing allows more airflow and light to come through and can serve as a good trellis for vines, but it can pose a risk if the openings are too large and a chicken can squeeze through or another animal can enter. A good compromise for an open design is to use lattice in areas where you may want more light and air circulation.

Fencing Zones

Different zones may benefit from different fencing.

○ *Perimeter fencing.* Perimeter fencing is a permanent fence structure that encloses an area, such as your backyard, your vegetable garden, or your patio and outdoor living area, to prevent or restrict access. In urban landscapes, property fencing is often a fully enclosed privacy fence that restricts access both physically and visually, to help prevent predators from entering, and to minimize the chickens' curiosity about what may be beyond. Because the perimeter fence is the first line of defense from predators, it's important to make this fence a high priority from construction and material standpoints.

> ### tip
>
> *Chickens are not likely to go over a fence without a firm or solid top rail to perch on as they fly over it. If a fence has slack or is unstable at the top, they may avoid trying to fly over it.*

○ *Cross-fencing.* Within the perimeter fence, cross-fencing is a good agricultural pasture-management practice that allows specific areas to be grazed while other areas recover. For chickens, the paddocks method can be applied to the landscape to allow the chickens' grazing to be rotated among several areas of your garden.

○ *Temporary fencing.* The simplest solution may be temporary fencing, if an area is in transition or your seasonal routines and rotations have not been fully identified. This kind of fencing requires only a few, reusable materials. Simple bamboo, metal, or fiberglass poles can be utilized in areas with bird netting or chicken wire; the poles can be laced through the fencing. Because chickens can't grab on to the top of this kind of fence to perch, they usually will not try to fly over it. Since most temporary fencing is not very secure, it should only be used within a predator-proof fence.

Regardless of the fencing method, if the chickens are kept out in the open for long periods of time, they will need plants or a structure to protect them from the elements and from predators.

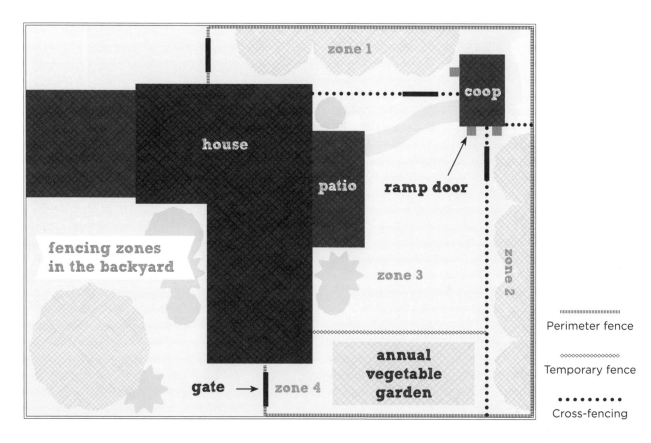

Electric Fencing

Hot-wire fences discourage animals from entering or leaving a space by zapping them with harmless but uncomfortable low-voltage electricity if they touch the wire. If a single wire is strung along the base of your fencing at approximately 6 inches off above the ground, it will help prevent burrowing or digging predators from getting access to the chickens. A hot wire placed across the top of a fence will prevent predators from climbing over the fence. Electric netting is also available and comes in many heights, but be sure to keep it tangle free or you will have a heck of a time getting it straightened out. Also, chickens can fly over it and if you don't keep it charged all the time, predators can chew through it. Some things to know about electric fencing are:

○ These systems need electricity. Solar power may be one way to power these fence systems. Conventional electricity may not be reliable during power outages.

○ All electric fencing systems are regulated with a transformer-energizer that will help spread a certain amount of power to the fence. You must match the transformer to your fence's needs.

○ When electrical fencing comes in contact with vegetation, it will ground out, making it less effective. So be sure to keep vegetation managed near the fence.

Good Neighbor Fencing

Good fences make good neighbors? Personally I think it is the neighbor and not the fence. But there is a style of fence along a property line that is appealing on both sides: the good neighbor fence. During construction, the simplest way to build a wooden fence is to do it one sided, leaving the other side plain and often looking unfinished. The design and installation of a good neighbor fence can take more time and add material costs, but the effort can pay off in your neighbor's happiness and perhaps a willingness to split costs.

Green Fences

My favorite kind of fence is a living wall, or green fence. It is essentially a row of plants grown to make a thick hedge, creating a barrier like a fence. Living walls can be quite functional in creating habitat and cover for chickens, but they may not be predator-proof or keep chickens contained unless you use them in combination with conventional fencing. When a mixture of plants is used for an informal hedge, it is often called a hedgerow. The feature can create important wildlife habitat for birds and other small animals by providing shelter and food (berries or nuts), as well as creating bee forage, and can also act as a windbreak.

When choosing plants for a hedgerow, think about whether they can be used effectively in combination with another kind of fencing to protect your chickens; the overall mature height and width of the plants; the growing conditions (sun or shade) and the soil conditions (moisture and acidity) of your yard; how fast the plants will grow; their hardiness and disease resistance; and their seasonal attributes such as color, fruit, and evergreen versus deciduous.

Plants for hedgerows or green walls

○ Arborvitae (*Thuja* species)
○ Bamboo (*Phyllostachys* species)
○ Barberry (*Berberis* species)
○ Boxwood (*Buxus* species)
○ Dogwood (*Cornus* species)
○ Elderberry (*Sambucus* species)
○ Hawthorn (*Crataegus* species)
○ Hazelnut (*Corylus* species)
○ Hemlock (*Tsuga* species)
○ Holly (*Ilex* species)
○ Mahonia (*Mahonia* species)
○ Osmanthus (*Osmanthus* species)
○ Privet (*Ligustrum* species)
○ Rose (*Rosa* species)
○ Serviceberry (*Amelanchier* species)
○ Viburnum (*Viburnum* species)
○ Willow (*Salix* species)

Existing Fences

In many situations, chicken owners will have to work with fences put up by a previous property owner. In that case, inspect the fence for wear and damage and make sure it will be effective for chickens. If a wooden fence is rotting along the ground, that part can be cut out and replaced with a new board. Chicken wire or wire mesh nailed to the fence along the ground is commonly used, but because it rusts in a few years and perhaps sooner if in contact with the ground, check it regularly. If an existing fence is in good condition but is just hideous to look at, a coat of paint or stain can make a big difference.

OPPOSITE: A coop and tall privacy fence in a chicken garden.

The Chicken Moat

A handy device for containing chickens is the chicken moat. These short, tunnel-like fences made from bent wire mesh along the perimeter of fence lines allow chickens to get from one area to another without having access to the space beyond. If you have a large edible garden, a chicken moat could be placed around the perimeter, so the chickens can patrol to keep out any incoming pests like grasshoppers or snails.

Gates

When you have fencing, you will have gates. With chickens, you are likely to need many gates, depending on your garden layout. When creating an entranceway into a garden area, you can be creative and make a statement. Or you can keep it as simple and functional as rolled-up wire fencing, if that is your preference. Typical garden gates are 3 to 4 feet wide, allowing most wheelbarrows or large garden tools like lawnmowers to pass through. If you need to have an area that opens for wider access, make sure the posts are strong enough to support the heavier gate. Another important factor is which way your gate will swing open, since it can block the best access if it opens in the wrong direction. Gates usually consist of three basic elements:

o *Hinges.* Gate hinges should be made of materials that will withstand outdoor weather. They should have a spring-loaded element that shuts the gate behind you, which can cost a bit more, but will be well worth it, as a deterrent to sneaky intruders or escaping chickens. A simple, effective, cheaper device to automatically close a gate is a bungee cord: place it on the fence so that it is stretched enough to pull the gate closed.

o *Latches.* Different gate locking mechanisms can range from a simple gravity latch or hook to a complicated system that locks with different levers. Keep in mind that some predators like raccoons can easily open gate latches. Many locking mechanisms come with the option to add a padlock.

o *Gate material.* A gate is usually made out of the same material as the fence. Heavier gates will most likely need some kind of anti-sag hardware, which can be as simple as a heavy wire spanning diagonally from the top hinge corner to the lower opposite side of the gate. Kits are available at hardware stores, and agricultural or feed stores will have farming hardware, which can be heavy duty and less ornamental. Prefabricated gates are another option. If you plan to use salvaged material, be sure it is weatherproof, since garden gates are heavily used.

Fencing and the Law

It's not uncommon for legal issues to arise over fences. Many jurisdictions have zoning regulations that may require fence permits, and that limit certain design aspects of fences along your property lines, such as height and distance from easements. Legal disputes with neighbors over property lines are prevalent, so be sure to find out exactly where your official property boundaries are, and talk with your neighbors before constructing fences along property lines.

CHICKEN TRAINING AND ACCLIMATION

Yes, chickens can be trained! It is easy to teach chickens simple commands; it may take some time and commitment, but it is worth it. I was fortunate to attend a Poultry in Motion workshop, in which Terry Ryan, a world-renowned dog trainer and author, has been using chickens for years to train professional dog trainers to improve their skills. Chickens are quick learners and force trainers to stay on their toes, making them an ideal training subject. After several days in this training program, the student's chickens can navigate through a series of ladders and bridges much like a complicated dog agility course. It is an impressive sight to see. In the garden, we don't need that level of training, although it would be fun. I have daydreamed about teaching chickens to deadhead spent flowers.

Animals have different reasons for learning from humans. And in most training techniques, the animal has to work for something. For dogs, it may be for praise or to please their pack leader or master. For chickens, they will simply work for food. When the animal does something we consider good, it can be rewarded. The reward can be marked by a cue, such as a sound like a clicker, a whistle, or a word. This process to teaching animals to do things we want them to do is called shaping behavior, and is very useful so you can easily catch your birds or call them to you when there is potential danger.

Teaching Commands

You will have good success with free-range chickens if you can teach them some basic commands. If your chickens are cooped up and you have decided to free range them, you may have some work ahead of you, compared to your having started your acclimation and training with young chicks. Keep in mind that chickens are sensitive and easily stressed, so be sure to work with them in a stress-free environment. Some useful commands to teach your chickens are:

○ *The "come" command.* Teaching a chicken to come to you when called is relatively easy but will take some time. It simply involves making the chicken know that you are safe, and this will involve, at first, giving it food. Trusting humans doesn't come easily to all chickens, and some breeds are more prone to being flighty than others, so those breeds may need extra handling. A chicken that has a full crop may not be as interested in treats you have to offer, so begin this training with hungry birds. Simply let them approach you when you have food in your hands or in a container. Once you get them eating out of your hands, make a noise that will cue them to associate it with your giving them food. I put the food in an old coffee tin and shake the tin so they can hear it, and I call out a high-pitched, "Hey, ladieeees." You can use a clicking noise, a whistle, or something else to cue them, but just be careful it is not a noise you would use at another time. Do this daily for several days, and eventually you will no longer need to bait them with treats; they will just come to you when you cue them. My chickens are conditioned to come to me every time I go outside or when I come home: they hear my car and come running. If I am outside gardening, there is usually a chicken or two hanging out with me, and the others are off on their own but will come running if I call them.

○ *The "sit" command.* This command works only with hens and is based on the fact that hens are submissive and will lower their body into a mating position when they are ready for the rooster to mount them. They lower their head, lift their tail, and sometimes start to stomp their feet in place as if they are trying to keep their balance. If there is no rooster in your flock, then they will naturally be submissive to you, and it is easy to get them to do this every time you need to pick them up. Once you see a hen squat for you, pet her back and let her know that you are safe. You can choose to make a sound or noise to associate with this behavior, but bodily language will probably be adequate. I hold up one arm over the hen, and she will squat. I will sometimes say, "good sit," but most of the time hens will do this even if not cued. Watch out, because it can be easy to step on a hen if you are in a full walking stride and she decides to "sit" for you.

Teach children how to handle the birds carefully. There are many ways to hold a chicken, but if you cup your hands over the chicken's wings when you pick them up it will make it easy and less stressful to them. It is easiest to hold them like you would a football.

You can also teach your chickens to do tricks. With time and some treats, you can teach them to do silly things like jump up and down or fly onto your arm. I try not to encourage my chickens to fly at me or onto anything, because it increases the chance of my getting pooped on or their flying over a fence. To hypnotize your chicken, simply hold its body down and head toward the ground and draw a straight line from its beak outward repeatedly, and the chicken will be in a trance for up to one minute. For other tricks you can teach your birds, see posted videos on the Internet. Some chickens respond better to training than others, and some do not appreciate being messed with. Chickens have egos too.

BELOW: Children can most easily hold a chicken like they would a football.

Acclimating Chickens to the Garden

When introducing animals (or children, for that matter) to your garden, there needs to be a period of acclimation or adjustment in which everyone involved learns the rules and the routine. If you just let your chickens loose in the garden, they are not going to know how to be well behaved. I've seen countless chickens that have been cooped up their entire lives get loose and completely destroy a garden in a few hours. It is like letting a kid loose in a candy store—they go crazy. So here's the procedure I recommend:

○ *Introducing chickens to the garden.* During the introduction period, you should slowly acclimate your chickens into the garden. Spend a couple of hours a day for a few weeks, and allow ample time for you to supervise the birds. Start in the late afternoon or evening, when the chickens are hungry enough to come to you when you offer food. Most chickens will return to the coop at night, but be ready with food in case they don't go in easily on their own. Over time, slowly add time to their outings and less supervision as you learn their habits and behaviors. You will be able to note which plants the chickens particularly favor and make efforts to protect any plants at risk. As the chickens get familiar with their surroundings, they will use plants as shelter and learn to duck under cover when they hear overhead noises such as hawks, crows, and even airplanes. This time to adjust will help them become adapted in the same way they would if they lived in the wild.

○ *Establishing a routine.* Everyone has a different routine with their chickens. I allow my birds to free range part time, and they work their way back to the coop. If you are raising egg layers and would like them to learn to lay eggs in the coop nesting boxes, keep them locked inside the coop during the morning, which is when they usually lay eggs. If the chickens have established a laying routine in their boxes and you keep the coop doors open for them to access, it is usually safe to let them out in the mornings. Most people have problems with younger hens who are just starting to lay, but if you have older hens laying in boxes, the younger birds will usually follow suit.

○ *Some chickens need a "time out."* Chickens learn from each other. While an older bird can teach younger birds the ropes, the young birds can also learn bad behavior from older or high-ranking chickens. Look for chicken behaviors that can be destructive to either the garden or to each other, and have a way to isolate the naughty bird if needed either in an enclosed run or maybe in a chicken tractor.

Chemical Dangers

If your chickens are free ranging near areas where chemicals are stored or used, be aware that materials you may use every day and think are safe may not be safe for your birds. Know exactly what you are putting into your garden, because your chickens will ingest fertilizers, slug baits, pesticides (herbicides, fungicides, insecticides), and coated seeds that can make them sick or even kill them.

tip

Be selective about the treats you give your chickens. Even if it is your leftovers, don't feed your chickens the hard-earned vegetables from your garden if you don't want them to eat them when they are out on their own. Use a worm bin for those food scraps, and then feed your chickens the worms.

to clip a chicken's wings, or not?

Whether to clip your chicken's wings is one of the difficult decisions you may face when owning chickens. On one hand, clipping a chicken's wings doesn't hurt them and will keep them from flying into or out of fenced areas. But on the other hand, it will significantly reduce their chances of escape in the event of a predator attack. Many chickens will not bother flying over fences if they can't see the other side or if they are too heavy. By clipping the primary flight feathers on one side of the chicken, the bird cannot fly because it will be off balance. If you do this with a young pullet, it can possibly inhibit the interest to attempt flying later as an adult bird. It's easy to clip the feathers, but it's best to do it with the help of another person:

1. Have your helper hold the chicken upside down and it will be subdued because the blood rushes to its brain.
2. Hold out one chicken wing so that the flight feathers are spread out like a fan.
3. Using sharp scissors, cut the primary flight feathers, which are approximately the 10 from the wing tip inward.

While the chicken is molting, you may need to help pull those cut feathers out. The flight feathers will regrow, so you will have to trim them again after the chicken molts each year.

designing a
CHICKEN-FRIENDLY GARDEN

A well-designed landscape can transform your garden into an extension of your home, offering an attractive and productive outdoor living space. For our feathered friends, a well-designed garden can offer protection and safe shelter as well as food for them to forage. While landscape design is probably not the first thing you think of when getting ready to bring home some chicks, it is an important consideration for raising birds that will live in your garden—especially if you want them to be free ranging. Whether you are starting with a blank slate or working with a fully established mature landscape, there are several steps to take in planning your chicken-friendly garden.

the
fries garden

THE BEST WAY I CAN DESCRIBE THIS GARDEN MASTERPIECE is to call it an enchanted wonderland—with poultry. Kathy Fries has taken a long, narrow, steep, suburban lot and transformed it into a space where anyone could get lost in an adventure of discovery. There are winding paths up the hillside and with every turn there is a new surprise, whether it is a carefully placed ornament or the entrance to a new lush garden room.

Her garden art has a whimsical air about it, much of it is chicken related, and large structures in her garden are all renovated from existing buildings, the coop included. What was a two-walled shed has been transformed into a chicken coop out of a fairy tale. The Palais des Poulets, nicknamed the Clucking Hen Palace, boasts a large tower that stands next to the entrance of the coop with an adjacent formal knot garden. An attached covered run looks like a fancy bird atrium, and has enough chicken-friendly features to make it, alone, a perfect chicken playground. The interior of the coop is half coop and nest boxes and half storage for chicken and garden supplies.

The Fries family has raised chickens for over eleven years and has had many rare heritage breeds. The birds are all hand raised inside of their home with their two young boys, Xander and Jasper, then they are moved into the coop when they are old enough to hang with the rest of the birds. Their flock has seen loss from predation both from the sky and ground: living near a lake, they are in the direct flight path of bald eagles that fish at the lake and fly over the Fries' garden to their nearby nest. A bald eagle once scooped up their largest rooster off the driveway, and since then they have provided more protection for the chickens. Kathy brings the chickens out into the garden but only a couple at a time when she is working in her edibles. She has successfully used the birds' manure and bedding for amendment to her garden beds and is now pasturing the chickens in an old orchard.

CREATING A PLAN

To begin, you'll need a base plan of your garden that shows the house footprint and property lines. There are resources where you can get this information without having to do all the measuring yourself. A good place to start is the Internet. Often there are online county records of your property. Not all counties have this information available online, but you'll want to check. Another place to look is in the paperwork from when your house was purchased. Often the appraisal documents include a drawing of the lot plan and measurements. You can also check out online satellite imagery. Especially in more populated areas, mapping websites will have bird's-eye satellite views of the land. If none of these options is fruitful, you'll want to get out the measuring tape and head outside to start drawing your base plan.

Draw all the existing elements that are going to stay.

On a large piece of paper and with pencil, draw the property lines, the footprint of the house and any out-buildings, and all the existing trees or shrubs that are going to stay. Include all measurements and make the drawing close to scale.

Create a wish list of the elements you would like to include in your plan.

On a separate piece of paper, list everything you want to include in your garden plan, such as the coop, the run, an orchard, and a raised vegetable bed. Your list may be long or short, but list everything you want to include and eliminate as you go, rather than trying to add an element to the design later. This wish list will help you decide how to assign and divide up your space.

Besides the essential chicken infrastructure, other items you might want on your wish list are a children's play area, sandbox, entertainment or patio space, fire pit, greenhouse, and hammock area. Browse garden magazines, books, and websites for ideas, to help you prioritize what is most important to you, your lifestyle, and your family's needs.

Chart your yard's sun exposure. It's important to know the exposure in your yard, because it will help determine where you site elements in the garden and what plants you choose. Start by locating north. Figure out where the sun rises in the morning and where

it sets at dusk. Throughout the seasons, these locations will change on the horizon. The sun creates different light and shadow patterns during different times of the year. An area that is in full shade all winter long may get direct sunlight all day in summer, dramatically changing what you place there. Also, your deciduous trees will let much more light through them in the winter, when their leaves have dropped, than they are leafed out in the summer.

I like to use colored pencils to shade in these different exposures:

o *Full sun.* A full-sun location should get 6 hours or more of direct sunlight during the growing season.

o *Partial sun and shade.* An area with partial sun and shade will get 4 to 6 hours of direct sunlight during the day.

o *Full shade.* An area of full shade gets 4 hours or less of direct sunlight during the growing season.

The growing season can usually be defined by the time of year between the last frost in the spring and the first frost in the fall. It varies in different climates (both micro and macroclimates) throughout any region and can be caused by environmental factors such as elevation, temperature, and rainfall. To locate your climate zone, check the table on page 204 which shows USDA plant hardiness zones based on average minimum winter temperatures. A map of the zones is also available online. I live in a Zone 7 microclimate: it is cooler than the nearby cities, which have an urban heat island effect and are Zone 8. The warmer the climate, the higher the USDA zone number; for example, areas near Anchorage, Alaska, are Zone 4 and those near Miami, Florida, are Zone 10.

Note the microclimate(s) in your landscape.
Are there areas in your backyard that are hotter or colder than others? Are there areas that are lower or higher in elevation, where wind would affect the plants? Are you close to a large body of water or in the northern shadow of a large hill or a tall building?

Mark any areas of concern. For example, is there a nasty view that you would love to block or a peering neighbor who makes you want more privacy? Is there an area that collects standing water during parts of the year?

site analysis

○ = Full sun
◐ = Partial sun
● = Full shade

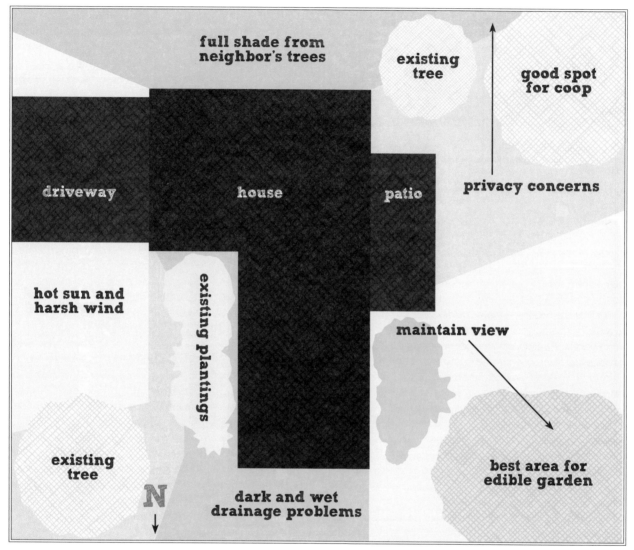

full shade from
neighbor's trees

existing
tree

good spot
for coop

driveway

house

patio

privacy concerns

hot sun and
harsh wind

existing plantings

maintain view

existing
tree

N

dark and wet
drainage problems

best area for
edible garden

Draw in major functional garden features.

In mapping out how you want to use the space and prioritizing use of the land, first identify what major functional features dictate where you place other items. These features are categorized as hardscape, or non-living, and softscape, or living, items. You need to place workable paths, for example, for circulating through the space; plan around where resources, like water, are that will be used regularly; and utilize the exposures for growing certain kinds of plants. Using your base plan, create a bubble diagram that sites these items. First locate the features that require a particular exposure, and then work around those. Refer to your wish list, and use the process of elimination in deciding what you can create in your space. Identify any leftover space, where beds and lawn could be placed. Draw lightly with pencil so you can easily erase what doesn't end up working for you.

You should now have a workable base plan. Good design doesn't happen overnight. Take time to think through these elements, and you'll arrive at a successful plan for your space.

THE CHICKEN INFRASTRUCTURE

Planning the infrastructure for the chickens is a significant part of your garden design. And placing the chicken coop on your site may be one of the most important decisions in your garden plan.

The Chicken Coop

As a structure, the chicken coop is the first of the hardscape items that you will usually place on your plan. Your chickens will be more comfortable if their coop is positioned well, your neighbors will appreciate it, and you will have an easier time working in and around the coop. Following are some considerations in picking out a spot for the coop:

o *The law.* First, check your community, city, or county code to learn whether there are any restrictions on placing animal housing on your property. You don't want to go through the effort and expense of building a chicken coop and then find that the city (or the neighborhood association) will make you move it 4 feet over. And don't assume no one is watching, because neighbors you've never met down the street could register a complaint. Some cities require that the coop be a specific distance away from any houses, your own and your neighbors', which may narrow down where you can put it. The code sometimes requires a chicken coop to be located 10 feet or more from all buildings, property lines, and streets.

o *Exposure and climate.* You want your ladies to have a cozy spot to call home, located where they get plenty of natural light but also where they are protected from harsh weather conditions like heat, cold, or wind. Chicken coops should not be placed in areas where water will collect, such as at the bottom of a hill, or in areas with saturated soils. Ideally, the soil in the coop area should be well drained and receive direct sunlight for at least part of the day. Areas that are too exposed could get cold winds in the winter months, and could get too much heat during hot times of the year. Having a shady retreat is important for chickens during summer months. By simply placing a deciduous tree near the coop, they will benefit from the leaf coverage in the hot summer months but will get light in the winter when the leaves have fallen.

o *Accessibility.* Some chicken keepers choose to site their coops close to the house where they can have easy access, and others tuck them away into the back corner of the garden. Just be sure that you place your chicken coop where it is easily accessible for you. You don't want to put it too far away, so it is out of sight and out of mind; the coop could easily be neglected if you are not reminded of it daily. Amenities like running water and power may be huge factors in where you site the coop, because running back and forth with heavy water buckets can make chicken chores a daunting task. For electrical access, it is always nice to plug into an existing outlet rather than create a new one or use extension cords, so if there is a plug on the outside of an existing building, try to utilize that.

The Chicken Run

A chicken run is the permanent or temporary area usually adjacent to the coop that gives the chickens outdoor access to fresh air, sunlight, and earth. In many backyards, it is a permanent area that is undercover and has deep mulch bedding. It may be a moveable temporary fence system, with portals and tunnels to get chickens from their coop to the run area, which you rotate throughout the garden. Or it may be multiple runs that are divided by cross-fencing that allow you to rotate the chickens through different areas at different times of the year.

Chicken Paddocks

If you choose this system, it's best to equally divide a space into four or five areas with cross-fencing for runs, so you can rotate the chickens through the garden at different times of the year. This method allows one fenced area to recover after chickens have grazed in it for an extended period of time. Different types of plants can be grown within the paddocks, and if planned well, certain crops can be grown so that the chickens are grazing at times of the year when the crop is least sensitive. For example, blueberries or raspberries are a plant that chickens can graze underneath and around for most of the year, but during harvest in the summer,

tip

In small urban lots, if local ordinances permit, consider building a run in your side yard where the space is not large enough for much else.

a deep mulch bedding system

Also known as a deep litter system, a deep mulch bedding system creates compost on the floor of your chicken's housing or covered run. This sustainable approach is practiced by many chicken farmers today. Several different bedding materials can be used. White wood chips and shavings work well, but be careful of some sawdust that has fine, dusty particles. Chopped straw, hay, shredded paper, and leaves also can be used as good bedding material.

You begin by placing a fresh layer of bedding approximately 5 to 10 inches deep in the coop and covered run area. When the bedding material is decomposed or broken down to half its original depth, you add more bedding. Depending on how many chickens you keep and their space, adding more bedding is done about once a week. This layering allows the composting process to begin, and the chickens keep the bedding aerated by scratching. This system requires less cleaning time, and may only require two or three cleanout times per year.

This system works best when you have an earth floor in the coop and covered run because the microbes can move freely from the soil through the bedding. Other floor types can use this system as well, but will most likely require more cleanouts.

A deep mulch system in a chicken tractor can be used to build soil and essentially make raised beds out of the litter in the chicken tractor. Once the mulch layers are at a certain height, the tractor can be moved to a different area, and you will have a compost pile that can be used for a raised bed once it finishes the composting process.

AN IMPORTANT NOTE: *The bedding has no smell as long as it stays dry. Be careful that the material you choose does not trap a lot of moisture, which can grow mold and bad bacteria that can make chickens sick, and make the chicken coop smell.*

you can put the chickens in other paddocks with plants that you are not harvesting.

The plant selection for your paddocks will largely depend on your growing season and how many chickens you keep. As long as there is some sort of forage and plants for shelter, your chickens should be happy and safe in a confined-range paddock system.

Permanent Fences

Drawing possible fencing lines out on paper and stringing up some lines are good ways to visualize fence structures before actual building. Also, if you are unsure where permanent fence lines should be, consider using temporary fencing until you figure it out. Perimeter fencing along property lines should be straightforward enough, but when you start putting in cross-fencing you need to plan for gates and good circulation through the space.

The Compost Area

In some communities, you can simply put chicken manure and bedding into a green recycling bin and let the utility company haul it away with the rest of your green debris. But for gardeners, it is wasteful to discard such rich material. Chicken manure is one of the major benefits of keeping chickens, so planning a composting system should be a priority.

The size of your compost area will depend on a few factors, such as: How many chickens do you keep? What kind of bedding system do you use? Do you have a lot of other green materials to add to the compost pile? At what times of year do you add that biomass? Most urban or suburban residential lots need one or two composting bins approximately 3 feet by 3 feet by 3 feet, or 1 cubic yard. I prefer to use two bins so I can rotate the material from one to the other. There are many ways to compost and many bin styles you can use, whether it is purchased or a do-it-yourself system.

PREVIOUS PAGE: The coop is sited well in Jennifer Carlson's garden—close to the compost and near the shed. OPPOSITE: Provide snacks in moderation; a chicken forages in fall leaves.

The closer the compost area is to the coop, the easier it will be for you to use. If it is on the other side of the garden and less convenient to get to, it's less likely that you will use or maintain it on a regular basis. Some considerations in placing your compost system are:

○ Check local regulations for any restrictions.

○ Compost bins should be placed in an area that does not collect water such as a low spot or an area with poorly drained soil.

○ Compost bins should not be placed directly up against a fence, a tree, or against any wooden structures, because this can rot the wood and make access difficult. By allowing the air to flow all the way around the compost bin, you will also have more uniform compost.

○ Consider the rainfall and climate. If the compost bin is near an eave that directs rainwater, it can lead to saturation and anaerobic conditions in the pile. If it is in full sun and exposed to wind, the compost may dry out quickly and need additional water to keep it moist.

○ Most composting systems are not pretty and can be easily camouflaged. Consider the angles at which the composting area is visible: Is the system adjacent to the patio where you entertain? If that is your only option, you may want to consider concealing the system with a screen or plantings.

tip

If you cannot easily visualize a plan on paper, consider using objects to mark the locations and boundaries of your garden layout. You might want to use 2 × 4s or even measures of string to mark out your chicken coop before settling on the location. You can also lay out ropes or garden hoses to plan pathways or garden bed lines, to get a sense of how the plan will look and work for you.

OTHER PERMANENT ELEMENTS

The next step in planning your chicken garden is to identify the hardscape items that are the bones of your garden plan, such as a patio, barbeque or outdoor kitchen areas, entertainment space, pergolas, raised beds, large stones, water features, and a fire pit. These items should be located in the garden where space and access allows. This important part of the design should be completed before you select any new plants. For example, the lines for pathways will determine how large your beds are, and in those beds there is a specified amount of space for plants. If there are plants to work around, then make sure you allow enough room for their growth when adding any new hardscape items to the area. Budget also plays a big part in hardscaping, and often these items are not easy do-it-yourself projects. When these items are sited, then it is time to work on other layers in your garden.

Pathways

Circulation is a key element in good landscape design: it will expedite your movement from one spot to another in your garden. Sometimes these lines will be straightforward and direct; other times, you can be creative and let the pathways meander and curve. Examples of major pathways needed in the chicken garden are from the house to the coop and from the coop to the compost area. Carefully plan the material of pathways in locations that have heavy foot traffic; for example, if a well-traveled pathway is lawn, you may end up with mud.

Lawn Lines

If you are going to allow your chickens to free range, then lawn can be a good food source for them. For ease of maintenance and as a good design rule of thumb, lawns should be a continuous line or edge that flows through the space and isn't chopped up by angles and objects. To make it feel natural, allow the lawn line to curve softly within the scale of the space: think about a mower being pushed along the edge. Lawn needs full sun and well-drained soil, and can rival edibles for the same space.

Plant Beds

Plant beds will probably be placed in the areas that remain after you have sited the structures and have placed elements like lawn or any special features like edible beds. Because plants come in different sizes and grow in different exposures and conditions, it is easy to place the beds by filling in the gaps after the hardscape components of your design are located. A good chicken garden should have ample bed space for plants that the chickens can use for shelter, but those plants can also be quite functional in other ways, be ornamental, and can soften lines in the garden, creating depth and seasonal interest.

CHOOSING THE RIGHT PLANTS

Our gardens are multidimensional, complex systems that can be good chicken habitat if designed well. Garden plants provide us with many benefits, like food, shade, and pleasure. To select plants to use in a chicken garden, I start by breaking down the layers in the garden and allotting spaces for those plants. For an ecologically sound chicken garden, the plants need to be well suited to the soil and climate conditions, and they should offer biodiversity and create habitat for wildlife and the beneficial insects and pollinators. The plants should not be disease prone or require unnecessary resources to keep them healthy and alive. Plant communities should work together with each other and the site where they are placed.

For chickens, a diverse plant community is imperative. Chickens need layers of plants to hide in and browse from. The more plant options you give them in your garden, the less damage they will do. And the fewer chickens you have, the less chance of problems with plants being decimated, so keep only as many chickens as you actually need. If you have well-established plants, the chickens will be less likely to eat them, and if they do take a taste, the plants will recover easily.

When choosing plants for a chicken garden, it is important to pick the "right plant for the right place," so the plants can thrive:

○ *Mature size.* Genetically, each type of plant has a size it will reach at maturity, so figure out how much space each plant will eventually need. Gardeners often attempt to control the size of plants by shearing or pruning, but some plants will suffer from this approach, leading to disease or pest attacks and possibly death.

○ *Hardiness.* Plants have a specific range of temperature that they like to grow in. Temperatures below a specific degree can kill plants, and if temperatures get too hot the plants will suffer. Know the hardiness zone in your area, and check that the plants you want will grow well there.

○ *Growing conditions.* Plants genetically need a certain type of soil and amount of sunlight. If you plant a sun-loving plant in full shade, it will not thrive and may become so stressed it will die; the same is true for soil conditions. Choose the right plants for your exposure and soil.

○ *Consider the scale of your garden.* While there are many elements to consider when designing a garden, such as form, rhythm, mass, texture, balance, and contrast, I'm only going to focus on scale, the most important factor in a chicken garden. Scale is when we look at the size of items in relation to one another. It is important to the human perspective because it is what makes us feel comfortable and fits the best in the amount of space we have.

food forest gardens

A food forest garden is a keystone concept in permaculture design. It mimics the natural relationships found in a forest ecosystem, by having many plant layers and offering biodiversity. A food or edible forest garden can be designed to be productive by not only growing food for us, creating habitat for wildlife and our chickens, and producing forage for beneficial insects, songbirds, and pollinators. Food forest systems can also grow fuel, fodder, fibers, and herbs that can be made into healing concoctions. There are several vegetation layers in a food forest garden:

○ Canopy, an overstory and understory of fruit and nut trees.
○ Shrubs that produce berries and seeds.
○ Herbs and vegetables, annuals and perennials.
○ Vines that produce food and are supported by the trees.
○ Groundcovers that yield food, prevent weed invasion, and reduce soil erosion.
○ Root and bulb crops, including mushrooms.

You can create a food forest garden in any aesthetic style in a typical urban, suburban, or rural backyard.

PREVIOUS PAGE: This chicken garden has many different layers: a deck, gravel pathways, planting beds, and eco-turf. OPPOSITE: These hens are foraging on a pathway that leads to a stream.

Trees

Trees are the first layer to consider. They will be the biggest and most permanent plants in your design and will give significant structure to the garden. Generally, chickens will not harm a tree by scratching unless it is very small. For new little trees, simply put temporary fencing around them until they get established.

Start by locating the areas where you want to have trees, keeping in mind the shade those trees will produce and how they will change the exposure in surrounding areas. Maybe you have an empty corner on one side of your lot and need a tree to screen your neighbor's hot tub, so consider an evergreen tree that provides some year-round screening. Maybe you want to have an orchard but don't want to lose any sunlight in the rest of the garden; consider planting fruit trees along the north edge of your property so the shade from those trees is cast to the north beyond your yard. Trees have many attributes that we may want or not, and so consider these characteristics when selecting trees. Are you looking for trees to bear fruit, have showy flowers, have fall color, or offer winter interest? Narrow these features down, and then find a tree that will work in the space that you have.

o *Conifers and evergreen trees.* Evergreen trees play a significant role in chicken habitat. Because leaves or needles are present on the tree throughout the entire year, the base of an evergreen tree can provide refuge for chickens and dryer soil during the winter months. My chickens' favorite place in my garden is under a western red cedar. They take cover there, and like to take their dust baths there. The wood and debris from this tree contain antifungal and insecticidal qualities, so it helps to prevent parasites in our flock.

o *Orchards and fruit trees.* Fruit trees make ideal chicken habitat, because chickens can forage in an orchard year round. Chickens help glean insects that may otherwise overwinter and infect the fruit of the tree. And any fallen fruit gets gobbled up by the chickens, making less food that you need to give them. Fruit trees come in many types and sizes, and many need another tree for pollination, so do your research. If you have limited room, consider espaliered trees.

o *Espaliered trees.* Espalier is a traditional horticultural practice of pruning and training a plant against a support to control woody growth. With this method, you can have fruit trees that take little space. And you can take advantage of favorable growing conditions,

like south-facing walls that absorb heat and offer an ideal microclimate for fruit trees. Espaliered trees take time to shape, but it's well worth it, and it is an ideal way to grow fruit in a small chicken garden.

mulberry

This deciduous fruiting tree (*Morus* species) is a terrific addition to a chicken garden. The tree is self-fertile, meaning you only need one to produce fruit, and it blooms in spring and then has delicious fruit that ripens all summer long. Mulberries are similar to blackberry but larger, and are great for making pies, jams, tossed into salads, and preserving by freezing or dehydrating. Mulberry also comes in species with white berries (*Morus alba*) and red berries (*Morus rubra*).

The berries are abundant and fall to the ground, where chickens can easily gobble them up, making it a great food source. Other wildlife love this tree's fruit, and it has been traditionally used by farmers as a decoy to keep animals and birds away from more valuable ripening crops. If you do not want wild critters harvesting your mulberries, you can try netting the tree or hanging pieces of reflector tape from the branches to deter them.

The mature size for mulberry trees varies for different cultivars; they can reach up to 60 feet tall and 30 feet wide. The tree can be easily kept smaller with pruning. Mulberry grows in full sun or partial shade, in many soil types. Keep in mind that this tree's dropped fruit can be messy, leaving stains on patios and swing sets, so place the tree in a spot where that won't matter.

Popular mulberry selections include: 'Illinois Everbearing', a cross between red and white mulberries, that is a grafted tree known for its vigor and hardiness (to -30°F); 'Sweet Lavender', 'White Mulberry', and 'Beautiful Day', which produce fruit that matures white and will not stain your fingers or patios; and 'Pendula', a weeping ornamental selection that makes a good hiding spot for chickens or kids, and is smaller than other varieties but can be staked to the desired height. Mulberry hardiness varies depending on the selection, but ranges between Zones 4 and 10.

OPPOSITE: Chickens eat apples that fell to the ground from a wall of espaliered trees bordering a vegetable garden.

Shrubs

Shrubs are the next grouping of plants to select for your chicken garden. Shrubs can provide aesthetic appeal, and can also serve as shelter and a food source for you and your hens. Shrubs come in many different sizes, shapes, seasonal colors, and flowers, and the options seem endless. You need to narrow down chosen shrubs to what will fit in your space. Shrubs are the backbone to mixed shrub borders, are perfect for foundation plantings and for filling a void in an empty area by adding bulk and depth to the bed space. Having a mixture of shrubs that are evergreen and deciduous will help ensure year-round interest and chicken shelter. Some shrubs can be pruned into tree form and make excellent small trees if you do not have space for larger trees. Most established woody shrubs will not be harmed by chickens scratching or dust bathing at their bases, but keep an eye on shallow-rooted shrubs that have foliage all the way to the ground. If chickens hang out underneath and dig holes, they can expose roots, causing drought stress.

Perennials

Hardy and herbaceous, perennial plants die back in the winter and return each spring, year after year. Perennials can add splashes of color at different times of the year and add unique textures to your garden. Some perennials are not ideal for the chicken garden because they can be easily damaged, while others seem bullet proof and aren't bothered by chickens or can recover quickly. Which plants will work best really depends on your garden, how you manage it, and how many chickens you have allowed to free range. Many perennials are easily divided, which means you can have more free plants later on, and most are easy to move if you messed up and put it in the wrong spot to start with. A few tips for using perennials in your chicken garden are:

○ Choose plants that are dependable and tough.

○ Choose plants that self seed or spread through rhizomes or vegetative growth to help ensure its survival.

○ In early spring as new foliage is emerging, you may want to have some form of protection ready whether it is a cloche, temporary fencing, or a teepee made of sticks.

blueberry

Known for their delicious and highly nutritious berries, this shrub is also a powerhouse in the garden, giving year-round interest with food, color, and minimal work. In the spring, blueberry bushes (*Vaccinium corymbosum*) display little pinkish flowers, develop fruit during the summer months, and in the fall they provide reddish fall color and reddish twigs for winter interest. Blueberries can be used in many ways—fresh, baked, and dried, and they are easy to freeze because they don't stick together.

Blueberries require soil with a low pH, which may require special amendments or fertilizers, and in hot dry summers they will most likely need supplemental water. Blueberries come in many different varieties that have different fruiting times, grow to be different sizes, and have different-sized fruit with different flavors. They are classified by heights into highbush, half high, low bush, and rabbit eye. Specific varieties do best in colder zones, while others do not need as much chill time and can be grown in warmer climates.

Chickens love blueberries, and when jumping up to pick off fruit one at a time, they can look like they are on a trampoline. Chicken keepers can either plant a lot of blueberry plants and share the bounty with the hens, or protect the fruit during the summer with simple bird netting (which I already do because of wild birds) and only give the birds the discarded fruit and let them clean up after harvest. You can also protect the ripening fruit by planting tough-leaved perennials or grasses underneath the shrub that chickens will hesitate to walk on. Or if this plant is used in a paddock, simply keep the chickens in a different paddock when the bushes are bearing. The taller varieties can be pruned into little tree forms, which makes the fruit harder for chickens to reach. I grow a variety of blueberries, all with different ripening times so the fruit is available all summer long.

Hardiness of blueberry shrubs depends on the variety and ranges between Zones 3 and 11.

OPPOSITE, TOP: Blueberry fruit can be protected under bird netting. OPPOSITE, BOTTOM LEFT TO RIGHT: Fruit is ripening on an 'Illinois Everbearing' mulberry tree, which is draped in Mylar strips to deter wild birds. Jerusalem artichoke.

jerusalem artichoke

Jerusalem artichoke (*Helianthus tuberosus*) is a flowering perennial that is in the daisy family and is a useful plant for us and for our chickens. Other common names include sunchoke, darth apple, sunroot, and topinambur.

A native in much of North America, it grows to be 5 to 10 feet tall, blooming in late summer or early fall with a bright yellow flower that is similar to a sunflower. It can be grown in many soil types including nutritionally poor soils, and prefers sun to part shade but not all shade. It is known for its edible tubers that look similar to ginger roots but are used much like a potato in recipes. The tubers store inulin carbohydrates instead of starch, making it a good food plant for diabetics. It can also cause gas when you are not used to eating it. The root can be fed to animals, and for chickens it is best to cook and mash the root, but chickens can and will also eat the leaves. This plant is considered a weed by some gardeners, since it will spread where it is happy. After it blooms, cut it down, and then, because the soil will be bare through winter and early spring in that spot, sow a winter grain or cover crop. A study has shown that Jerusalem artichoke stimulates the growth of broiler chickens and helps protect them from toxins and potential pathogens.

Ornamental Grasses

I love the soft texture that grasses bring to a landscape. They can be beautiful, useful plants, and there are many grasses to choose from for your garden. Clumping and running types of grasses grow in all different soil types, are low maintenance, and for the most part are disease and pest resistant. Low grasses make good borders along pathways or at the edges of any bed. Some grasses are compact and tidy in appearance, while others have a loose, wilder look. Large grasses have graceful movement and can be used much like shrubs for impact or as hedge material. In the late winter, you cut them down and can use the grass as bedding or nesting material for your chickens. Evergreen grasses along a border can act like raised edging, preventing chickens from scratching mulch out of the bed. Just watch out for plants getting buried by the chickens, some grasses will be fine with that while others will not. Many types of grasses make great cover for wildlife and chickens while also providing tasty flowering seed heads. Generally, chickens eat only low, soft ornamental grasses and do not cause irreversible damage.

variegated japanese sedge

The evergreen sedge *Carex* 'Ice Dance' is an attractive, low-growing, grasslike plant with striking white edges. Technically not a grass, it prefers moist, rich soil, and partial sun, but I have seen it thrive in full sun and full shade, and it is not as dark green in the sun. This plant spreads slowly underground, creating a dense mat, and is extremely durable with small animals running over it. The plant can be divided easily, and can be used as a good dense groundcover in forested areas. Growing in Zones 5–9, it might have some tattered leaves after a harsh winter, but they can be sheared off in early spring to allow for a new flush of growth. Plant it at the base of shrubs that you don't want chickens scratching around or eating the fruit of, such as blueberries.

lilyturf

Lilyturf (*Liriope* species) is a lush alternate evergreen grass plant that looks and acts much like *Carex* 'Ice Dance' but prefers drier conditions, making it a good drought-tolerant grass. Growing in Zones 4–9, it can be used as borders, groundcover, and even has little purple flowers in late summer followed by little black berries. There are two species: *Liriope muscari* is a clumping grass that comes in several nice varieties of different colors and sizes. *Liriope spicata* spreads by rhizomes, quickly if it is happy, so make sure it is what you want before planting it.

OPPOSITE, TOP TO BOTTOM: Variegated Japanese sedge. Lilyturf works well as a border in the garden.

Annuals

Most gardening chicken keepers will tell you not to bother planting annuals if you have free-ranging chickens. For the most part, I agree, because annuals can be easily destroyed if chickens eat them or scratch near them. While annuals are beautiful, they complete their life cycle in a single growing season, are delicate, and the money and effort spent grow them could easily go waste. But there are several annuals worth planting in your chicken garden that work quite well for many reasons: for example, if the annual is functional beyond aesthetics, such as being especially attractive to pollinators or beneficial insects or a natural repellant for garden pests. And I find that self-seeding annuals that return every year without much tending are especially worth trying.

nasturtium

A favorite annual in my garden, nasturtium (*Tropaeolum majus*) is a brightly flowered, cheerful plant that climbs throughout my garden beds and is a good companion plant. Loving full sun, it will creep quickly in and around beds and display flowers in oranges, reds, and yellows, and some varieties have variegated foliage. It is one of the easiest annuals to grow from seed and has so many uses. First, it is edible: we can eat the leaves and flowers, which are packed with vitamin C and make a great addition to a salad, adding a peppery taste. Animals can eat nasturtium too, but my chickens tend to steer clear of it. All parts of the plant have medicinal qualities and it has long been used as a healing herb. The seeds have a special vermifuge quality, and can be used as a preventive wormer for poultry. Because the seedlings of nasturtium are vulnerable to chicken scratching when they first come up in the soil, my trick is to plant them in a place that's hard to reach, and then the seedlings quickly grow beyond that spot once they are established. For example, grow them between a fence and post or at the base of a plant that is fully established and has some cover so the chickens don't go there.

Vines

The vertical nature of vines makes them a perfect addition to a chicken garden because chickens don't bother foliage that is high overhead. There are many types of vines for all situations, and some are heavy and vigorous while others are slow growing and dainty. Examples are kiwi (*Actinidia* species), honeysuckle (*Lonicera* species), akebia (*Akebia* species), and grape (*Vitis* species). Generally, in a chicken garden it is safer to grow plants that are tough. Many vines provide fruit, which drops and is then a treat for chickens. Certain edibles can be pruned and trained into vines, like tomatoes and squashes; and depending on the crop, the fruits may need added support. Vines can be planted on the outside of run fencing, and will grow up the fencing and drape overhead, providing shade in hot summer months. Be careful, however, not to use vines with poisonous foliage if the birds are cooped up full time. Vines can be grown on fences, trellises, arbors, and walls. The roots and trunk of vines should be protected from chickens as they are getting established.

Groundcovers

My chickens avoid entering areas that have a well-established groundcover layer. Groundcovers come in all sorts of heights and attributes, and help to keep your garden low-maintenance because they act like living mulch. They shade nearby plant roots, inhibit the growth of weeds, and provide erosion control. Groundcovers also minimize the need for buying mulch, since leaves and debris can fall to the ground and decompose with no cleanup needed. Groundcover plants that spread quickly have the best chance of survival in a chicken garden. Delicate herbaceous groundcovers can become easily damaged, but dense evergreen groundcovers work really well; for example, Japanese spurge (*Pachysandra terminalis*), ground raspberry (*Rubus calycinoides*), and cotoneaster (*Cotoneaster* species).

OPPOSITE, TOP TO BOTTOM: Nasturtium trails over a fence. Chickens avoid entering a garden with a mix of groundcovers like beach strawberry (*Fragaria chiloensis*) and sedums (*Sedum rupestre* 'Angelina', *S. spurium* 'Tricolor').

bamboo

Bamboos (for example, *Phyllostachys*, *Sasa*, *Arundinaria*, and *Fargesia* species) are one of the most useful and sustainable materials in the world. Many gardeners, however, might hear the word bamboo and have nightmares of an invasive plant taking over their garden. Technically a grass, bamboo has been used around the world for centuries as a textile for clothing, as well as for building material, food, medicine, musical instruments, and weapons. Bamboo grows quickly and can be harvested every season: the canes make ideal garden stakes. It is beautiful and functional in the chicken garden and there are many appropriate species available.

Bamboo can be an excellent evergreen habitat or cover plant, and because of its durability, it is nearly impossible for chickens to damage with a little pecking here and there. Another great feature of bamboo is that the leaves slightly flutter in a light breeze, making the planting a good sound barrier, which can be good if you have neighbors who might complain that your chickens are noisy. Bamboos can have two kinds of growth habits, runners and clumping. Species of bamboo that are runners can be contained with underground barriers to prevent them from taking over a space. Clumping bamboos are easily contained and come in a variety of heights. Some of the hardiest bamboos are *Fragesia* species, which are hardy to Zone 4. Most bamboos need full sun and plenty of water while they are getting established. Some dwarf bamboos are compact and can be used as an effective groundcover. Popular bamboos include:

○ *Fargesia rufa* 'Green Panda', a clumping bamboo that reaches 6 to 8 feet tall and grows well in Zones 5–9. It is vigorous, making it a good evergreen or hedge plant.

○ Black bamboo (*Phyllostachys nigra*), one of the most decorative bamboos with a striking ebony culm, or stem. Hardy to Zone 6, it can grow up to an impressive 30 feet with 2-inch-thick canes that start out green and then mature black. It is a runner form and needs to be contained; it also looks dramatic in containers.

○ Dwarf white-stripe bamboo (*Pleioblastus variegatus*), a running variety that will reach 4 feet maximum in height. It has beautiful contrasting leaves with bright white variegation, and makes a great groundcover. It tolerates both shade and sun, and is hardy in Zones 4–10.

Edibles

Edibles in a chicken garden can create frustrations. Because edibles take time and hard work, and we want to enjoy the harvest, we feel cheated if weather or animals destroy these precious plants. We can regard edibles in two ways: we can either completely separate our chickens from the edibles with fencing, or we can plant enough for all to enjoy and take as many precautions as possible to protect our crops. In my experience and that of many other chicken keepers, if there is enough food and space elsewhere in the garden, and chickens aren't raised with "treats" from the garden, they don't gravitate toward the edibles; but there needs to be plenty of other planting areas in the garden they can use.

In my own landscape, edible plants are integrated in with ornamental plants throughout, and I also have one area that gets fenced off in late summer when sensitive crops are ripening. I spend more time protecting my plants from squirrels and wild birds than from chickens, so I don't mind if my chickens help themselves occasionally. And I keep in mind that their rich, composted manure helps improve the soil so I can grow edibles.

Tips for growing edibles in the chicken garden include:

○ Have one designated area that can be fenced off either with permanent fencing or temporary fencing when the plants are most sensitive.

○ Use containers that are shaped so that chickens can't jump up on them and scratch or peck.

○ If you have free-range birds and don't want them to eat your tomatoes, then don't give them tomatoes as snacks.

○ Plant more than you need and share with the chickens.

➡ **CAUTION:** Always use barrier methods around low-growing edible plants that may be pooped on. Fresh manure on fresh lettuce will make you sick.

OPPOSITE, CLOCKWISE: A Bantam hen sits in a raised herb bed. A chicken seeks shelter under bamboo. A Buff Orpington hen seeks out pests in a vegetable garden.

EXTRA ELEMENTS FOR THE CHICKENS

Consider adding features in your garden just for your chickens to use, such as places to perch and a garden bed just for them.

○ *Water features.* With chickens free ranging in our garden, we need to have clean water available to them. A water feature can be beautiful and functional for chickens while also providing pleasant sounds and a habitat for other wildlife.

○ *A dust bath area.* Chickens will usually find a dust bath area on their own, which is often a dry area where you have a hard time growing plants successfully. If there is not a safe, dry soil area in your garden for chickens, create one. You can build a dust box for your chickens: it would be just like building a simple sandbox for kids, and it should have a lid or roof to keep the soil dry.

○ *Bug log.* A chicken's favorite food is any insect it can catch flying or crawling around. You can create a bug smorgasbord for your chickens by placing a log or wood round in a spot overnight, then flipping it over in the morning, much to the delight of your foraging chickens.

○ *Mirrors.* I've seen mirrors hanging in many chicken gardens for the hens to entertain themselves. Not only is it fun for the chickens, but a good-sized mirror can add the perception of depth in your garden.

OPPOSITE, TOP TO BOTTOM: A Plymouth Barred Rock hen drinks out of the pond. A simple shelter can be created as a dust bath area for your birds. BELOW, LEFT TO RIGHT: Two Buckeye hens eat insects from the ground where a log round sat overnight. A hen looks in a mirror placed at the bottom of a fence.

SAMPLE GARDEN PLANS

To help you visualize what your space could look like and how to lay it out, I have drawn three sample plans showing creative ways to have a garden and include chickens: an edible garden, a family garden, and a smaller garden.

An Edible Garden

The edible garden uses the paddock method for keeping chickens, with three different garden zones for rotation. Zone 1 is an orchard area with fruiting trees and shrubs, zone 2 is a cane fruit area, and zone 3 is a winter chicken range which uses tractoring on beds in other seasons. This design maximizes the space and exposure for edible food production.

edible garden with chicken paddocks

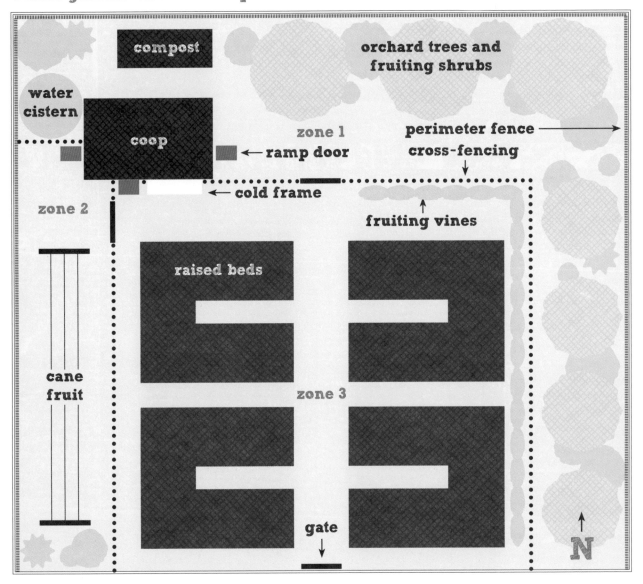

A Family Garden

In this family garden design, chickens have a permanent run outside their coop with shrubs for shelter, and they can free range part time in the rest of the garden. Next to the coop, a water cistern is used for collecting water from the roof.

family garden for part-time free-range chickens

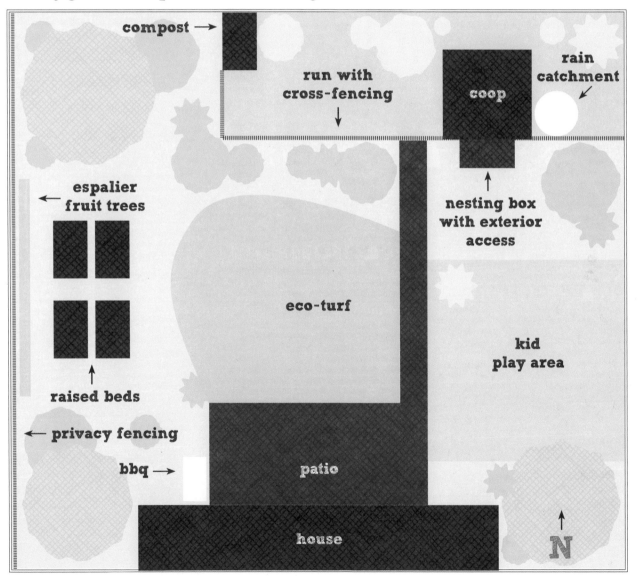

compost →

run with cross-fencing

coop

rain catchment

espalier fruit trees ←

nesting box with exterior access

eco-turf

kid play area

↑ raised beds

privacy fencing ←

bbq →

patio

house

N

A Smaller Chicken Garden

In small spaces, we might be more sensitive to our environment, noticing sounds and smells more than we might in larger spaces, because we have less distance away from neighbors and busy streets. When we have limited room, often we cannot choose large plants or create deep or dense borders, so it is important to choose plants and their placement wisely. Chicken keepers with small gardens can add specific plants to gain privacy from their neighbors, have pleasant fragrance, and also introduce sound to their space.

In this small urban backyard, a 6-foot-tall privacy fence surrounds the entire perimeter so the chickens can free range as the owner allows. The chicken coop has a green roof and an attached run with a clear covering that lets natural light in. The water from the roof is collected into a rain catchment cistern, for drinking water for the chickens. The coop has nest boxes that are easily accessible from the outside, making collecting eggs an easy task, and there are three nearby compost bins half buried in the ground. The holes drilled in the sides of the bins allow organisms in the soil to access the material and help to decompose it, and when one bin is full the next one can be filled. Following this rotation, the third could be ready for harvest. An eco-lawn in the center provides greens for the chickens to forage when they are out of the coop. Several key plants are repeated to provide privacy, and also offer sound with a slight breeze: tall grasses such as *Miscanthus*, quaking aspen that fill the corners nicely, and bamboo, which could be the clumping variety planted in the ground or a running variety kept from spreading by using barriers or large containers. A small water feature close to the patio adds an interesting focal point and a soothing sound. Miscellaneous vines, shrubs, and groundcovers are added for layering and fragrance.

OPPOSITE: This Buff Orpington happily ranges while children play nearby.

design for privacy, fragrance, and sound

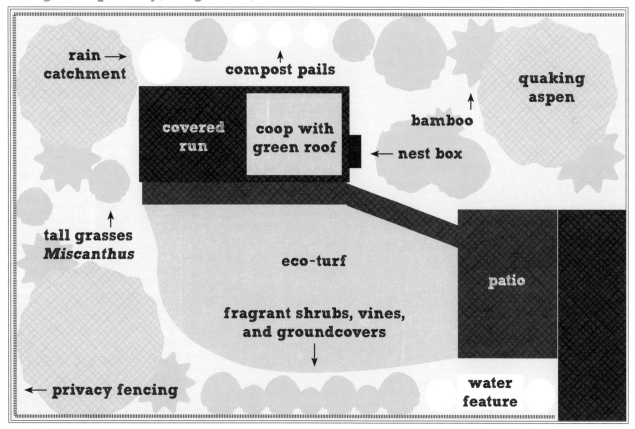

rain → catchment

compost pails

quaking aspen

covered run

coop with green roof

bamboo

← nest box

tall grasses *Miscanthus*

eco-turf

patio

fragrant shrubs, vines, and groundcovers

← privacy fencing

water feature

landscape materials

FOR CHICKEN GARDENS

Every garden is going to contain a mixture of materials, both softscape and hard-scape. If your garden already has materials in it, you will want to know what will work well with chickens, what will not, and why. Often, an established garden can be the best type of garden for chickens and will only need a few changes to accommodate the birds, whether it is removing some plants or adding some fencing. Mature plants offer many benefits: they aren't usually easily damaged by the chickens, and are great to work around.

the
hen poo garden

WHEN I FIRST APPROACHED ALANA MEYER'S GARDEN, I knew right away that there was something special about it. After all, how many gardens do you see on an urban lot, with the front yard showcasing corn? Alana is a certified professional horticulturalist and her husband is a blacksmith, and they live on less than one-quarter acre only two lots away from Interstate 5 in one of the largest cities in the state. They have five hens that free range all day in the garden.

Alana's approach to her garden is that there "should be enough food for everyone to share," including the birds, bees, humans, and other wildlife. Her natural approach even goes a step farther: when her hens become broody, she special-orders fertilized eggs so the broody hens can raise a few chicks each year. She doesn't believe in caged animals and fighting Mother Nature's natural processes, which also prevents her from adding more lighting to the coop in the winter, which would trick the chickens into laying because of what they sense is a longer day length. The short winter days are a time for her hens to rest from egg production.

Their coop is in the center of the garden along the fence between the garage and a small pond directly outside the kitchen door. Over the years, they have made several modifications to the coop, and are happy now that they can walk inside it and clean it easily. The roof is clear, which allows maximum natural daylight, and doubles as a drying rack for onions and garlic during harvest time. The nest box is one large covered area for multiple hens to use, is draped with fabric so it is dark inside, and is accessible from an outside hatch to gather the eggs. The garden is lush with diversity and layers, which Alana credits to her employment at a local nursery for many years. Beautiful metal art is displayed in all areas of the garden, even down to the smallest detail in gate hinges and coop hardware.

Plantings consist of a few small, well-placed trees that provide shelter and berries. Shrubs are ornamental as well as functional: over thirty blueberry bushes are scattered throughout the landscape, and the number is growing because she wants to be able to share more with the chickens. Alana's birds also work in the garden by helping her turn the compost. A simple metal grate is placed in front of the three-bin compost area, and when Alana needs it turned, she simply pulls the grate away and lets the chickens go to work. They help speed the process of breaking down the organic material by pecking through it for edibles and they aerate the pile by scratching. She then puts the material back into a pile, replaces the grate, and lets the composting process start again. Her horticultural knowledge and her relaxed attitude combine for a beautiful and successful free-range chicken garden.

OPPOSITE: Proud mama hens stroll with their chicks.
NEXT SPREAD, CLOCKWISE: Alana with one of her hens. Raised beds with cold frames are used to start seeds. A cobble path meanders through Alana's garden. A hen in Alana's garden. A Buckeye hen scratches in the gravel path lined with various edging materials.

IF YOU ARE STARTING A GARDEN DESIGN FROM SCRATCH and have your layout started, then you will have to pick out the materials and plants you want to use. Most new chicken owners will have some plants and structures in place, but will need to know how to maximize the space and make it chicken friendly.

HARDSCAPE

Our outdoor living areas are often comprised of patios or decks with pathways leading to and from areas. We need hardscape to keep down mud in spots that get a lot of traffic in the rain and for firm footing for our patio furniture or other outdoor items.

Commonly Used Materials

The materials and design options for hardscape can be endless, in aesthetics and budget considerations. Not all chicken owners are going to have the skills or tools to complete hardscape projects, and if that is true for you, then it is best to hire experienced professionals. Chickens generally will do well with the hardscape surfaces we live on and walk on, but there are some issues to consider.

Chickens leave behind their droppings wherever they go, and the manure can get stepped on and tracked around on your shoes if you don't clean it up regularly. Some chicken keepers keep a designated pooper scooper ready to clean up after their chickens, while others separate off the areas for humans from the chicken areas with low attractive fencing. My chickens tend to stay away from our outdoor living spaces, especially if we are having a BBQ or a lot of people are around, probably because they know they'll get their little cheeks pinched and don't want to be embarrassed.

You should consider the permeability of materials you choose in your hardscape areas and be sure that they have good drainage underneath and around the perimeter. With free-range chickens, it is likely that you will be hosing off manure from time to time. In warm or dry areas, manure hardens enough to be scooped up or swept off easily, but this would not be the case in areas with constant rainfall. Permeable materials are a wonderful low-impact development (LID) construction method, especially for urban areas where we have too much storm water runoff causing problems in our waterways. However, those porous surfaces or large gaps between surfaces can become clogged with debris, including manure.

○ *Stone.* Stone will add timeless beauty to your landscape. If you have sandset flagstone in random-shaped pieces and are trying to establish groundcover between the steppingstones, be ready for the chickens to scratch in the joints and potentially rip out the groundcover. You won't have that problem, though, if you mortar the material or have tight joints between the stones. You can also use the largest pieces available in the stone yard, or grow an aggressive groundcover and keep the chickens out of the area until it is established. Brass buttons (*Leptinella squalida*) is my favorite choice for a durable, fast-growing groundcover between steppingstones. Just keep the chickens away after the plants are first installed to allow time for them to put down roots and start spreading.

○ *Concrete.* Concrete comes in a wide array of poured and cast materials, colors, and patterns. One of the easiest and cheapest options is poured concrete, which many homeowners can tackle by themselves. But if poured concrete is not done correctly, it can cause drainage problems because it is impervious. If you have a concrete patio that you'd like to remove, then you can break it up, and chunks of concrete (also called urbanite) can be used for many garden projects, by you or by someone else. Concrete pavers and brick are among my favorite hardscape materials, because of cost, durability, design options, and ease of construction. They can also be hosed down easily.

OPPOSITE, TOP TO BOTTOM: A stone circle detail in the gravel pathway adds interest. Hens are foraging near an outdoor room with fireplace, patio, and pergola.

○ *Decking.* Decking comes in a wide variety of materials as well. Because wood is absorbent, it is best not to let chickens hang out on unsealed wood where their manure can get embedded in it. A good sealant or deck paint will help resolve that issue. The newer composite decking materials are another option, but they cost more and have different construction requirements.

○ *Gravel.* Gravel or crushed stone, of small particle sizes and any parent material, are my number one choice for pathways in a chicken garden. The birds can use it as a source of grit, which they need to break down food in their crop. Grit must be available to chickens, whether you purchase and provide it for them or they find it in nature as they forage. Grit can come from the earth in the form of sand, gravel, or mineral material available in bagged form such as granite rock dust. And if the chickens scratch in the gravel, they will help keep weeds down in pathways. It provides good traction, and it is the least expensive hardscape material to build with. It is durable and easy to install for just about anyone with some muscle and a strong wheelbarrow.

A few things to know about building gravel paths are:
- Gravel paths should be at least 3 inches deep.
- For a proper gravel base, you should dig out any organic soils until you reach a solid mineral subsoil, which might only be a few inches down.
- Use a plate compactor or similar tool to help the stone settle and become compact for use.
- Material should have smaller particulate matter, or rock dust, which helps bind the stones together, as opposed to clean or washed rock.
- I don't recommend using landscape fabric beneath the gravel.

○ *Wood chips.* Wood chips can be good low-cost pathway material, but it will need to be replenished over and over again with time as it breaks down. I particularly love using wood chips in a woodland setting or in areas with a lot of sensitive tree roots. Wood chips can sometimes be free if you ask a tree company or arborist to deliver them after a job, or they can be low cost as "hog fuel" which is a byproduct of land clearing, usually ground stumps and tree parts. Just be sure to ask for "clean or animal friendly" hog fuel if you are going to purchase it. Cedar playground chips can also be used. Cedar chips are safe to use with chickens as long as your chickens aren't confined in a closed space with it. Cedar also repels insects.

Irrigation Systems

Chickens don't care how you water your plants, but they may cause more damage to a stressed plant that is not being watered correctly. There are a few things to know about the different methods of watering and how they might be affected by your chickens.

Modern irrigation systems can be very high tech and are able to measure the moisture in the air and in the soil. Rain sensors know when it has rained and how much, then adjust their cycles accordingly. A smart irrigation system is not completely foolproof, and will cost a fair amount of money to install. Even a smart irrigation system can get out of tune, if it is not managed regularly, which can result in wasting water and money, eroding soil, and killing plants. Some of the irrigation systems you could use in a chicken garden are:

○ *Automated irrigation.* An automated irrigation system with pop-up heads that come on at scheduled times will not affect your chickens. They will typically run for cover from the spraying heads so they don't get drenched, but that's okay. These systems can be programmed to come on at certain times of the day, typically in the early morning hours.

○ *Soaker hoses.* Soaker hoses are a low-cost way to deliver water to your plants. There are some drawbacks to this method, though: if there are elevation changes in your garden, the low spots will get the water and the high spots will remain dry or not get an equal amount of water; and with chickens scratching and flinging mulch and soil around, the hoses can easily get buried and clogged. If using this type of watering system, plan around grade changes and watch for equal water distribution.

○ *Drip irrigation.* Drip irrigation is probably the most efficient watering system if it is designed and installed correctly. Above-ground drip emitters are placed at the base of each plant to deliver water directly to the roots of that plant. However, with the small feet of animals and children running through garden beds, these systems may fail easily. The tubing can be disconnected and leave the pressure of the system and water flow out of whack. Just like searching for that one broken bulb in a string of Christmas lights, a defunct drip system can be a chore to manage. Underground drip systems are good too, but for intense gardeners who are actively planting, dividing, and transplanting, be careful not to slice open an underground drip line with your shovel or trowel.

○ *Hand watering.* Watering by hand is my preferred choice out of all the watering methods, but is not practical for all gardeners because it takes a lot of time. Using this method, you can watch soil moisture levels and recent weather activity, and know exactly how much water your garden needs and when. Watering deeply and infrequently helps get your plants established and direct their roots downward, instead of at the surface of the soil, which makes them less tolerant to heat and drought. You can save enormous amounts of water, compared to what you would use with a system that comes on regularly or delivers water overhead, resulting in evaporation. A big downside to watering by hand is the amount of hose you need to drag through the garden; hoses can break plants, and they can be dragged through chicken poop.

Fencing Materials

Fencing comes in a wide variety of materials that differ in cost, function, and aesthetics. It is worthwhile to research exactly what you want, since your fencing must be effective in protecting your animals and your garden, and you will have to look at it every day.

○ *Wood.* Wood is the most common fencing material. You can purchase wood fencing at any local hardware store and it is relatively easy to build with. There are many grades of wood from different lumber, such as cedar, redwood, fir, and hemlock, and they have variable life spans and costs. Pressure-treated lumber is a common material used for outdoor projects because of its chemically assisted lifespan, but if you plan to grow edibles in close proximity to your fence, I caution against using this material because of toxicity concerns. Wood fencing can be painted or stained to prolong its lifespan, which can also create health concerns near edibles. But there are many environmentally friendly, approved paints on the market. And painting your fencing gives you an opportunity to introduce color into your garden.

○ *Metal.* Metal is another widely used fencing material. Common agricultural T-posts are a good option if you want a solid metal post for the long term, and they can be pounded into the ground with no digging required. There are many types of metal mesh fencing, in different patterns and gauges for varied uses. Most of them are galvanized, which are somewhat weather resistant, but some are PVC coated, which is more long-lasting and comes in a variety of colors.

○ *Chicken wire.* The most commonly used metal mesh material for chicken coops and runs is chicken wire (also known as poultry wire), which is an inexpensive, thin-gauge, twisted wire fencing with a hexagonal pattern of a small size intended to keep chickens in but allow air flow. Over time, it can rust and become brittle. It works well in the garden as a temporary fence to keep chickens out of certain areas. Because it is one of those materials that the chicken cannot grip firmly at the top to fly over it, it is a great choice for containing chickens, hence the name. But if your fence is multipurpose and you'd like it to support vines, chicken wire will need support to hold up any weight and will not be good for long-term use. Chicken wire may not keep predators from ripping it apart and getting to the chickens in a coop or permanent run.

○ *Hardware cloth.* Hardware cloth is another metal mesh option that is welded, readily available, and comes in a variety of small opening sizes. It is stronger than chicken wire but more expensive, depending on the gauge, length, and height you need. This material is my top choice for fencing material used for coops and permanent runs, where the chickens are locked in for long periods of time.

○ *Hog fencing.* Hog fencing is another common metal agricultural fencing material that is one of my favorites. The openings are usually 4-inch squares, or you can get it in a "combo fence," where the openings are smaller at the bottom. It is made of thicker gauge steel, which will last longer and can be used in a variety of situations

OPPOSITE, TOP LEFT TO RIGHT: A wooden gate is made of hog fencing and a layer of hardware cloth. Welded 2 by 4-inch galvanized wire is attached to pressure-treated wood. BOTTOM: This fence is not chicken proof but would deter a medium-sized dog. The rocks protect the plant.

with other animals and in the garden, but is not always predator proof. It can be used as a long-lasting trellis and is great for climbing plants.

o *Plastic fencing.* Usually made of vinyl or PVC, plastic fencing is available at low cost and in a variety of mesh sizes for privacy fencing, rail fencing, and more. It is touted as a low-maintenance material, but some plastic meshes can become brittle over time from the weather and end up just being dumped in landfills. Predators can easily chew through plastic fencing. I use plastic fencing only for temporary or portable fences within a sturdy perimeter fence.

o *Electric fencing.* Electric fencing can be very effective in keeping animals out of your garden or keeping your animals and birds in.

o *Masonry.* Masonry walls can work beautifully in the landscape. During the warm summer months, a stone or concrete wall can help ripen vegetables quickly by creating a heat sink if the wall is positioned to catch the sun. When properly built, they are virtually maintenance free and can be incredibly strong. The downside to masonry walls is first the cost of stone (unless found onsite or nearby), and the cost of hiring a builder with the skills to construct this type of fence correctly. Another downside is that many types of masonry are easy for some animals to climb over.

The effectiveness and costs of fencing materials vary depending on your region, the design, and the animals being protected and the predators being kept out.

tip

An opening of ¹⁄₂ inch is all a raccoon needs to grab a bird. When chickens are to be kept in permanent close quarters, be sure to keep this in mind when purchasing fence material.

OPPOSITE: Bird netting is easy to attach to other fencing, such as this birch log (not a living tree) used as a post.

fencing materials

MATERIAL	PROS	CONS
WOOD	Easy to obtain Easy to build with Easy to customize with style and color	Decomposes over time Top wood rails are easy for chickens to perch on
METAL WIRE MESH	Easy to build with Inexpensive Many mesh choices	Can rust and deteriorate Some can be ripped open by predators
PLASTIC	Inexpensive Easy to put up and take down in temporary situations	Can be chewed by predators Can become brittle over time
ELECTRIC	Highly effective Relatively low cost	Maintenance: Can short out if it touches vegetation Uses electricity — unless it is on solar, power, outages may be problematic
MASONRY	Sturdy	Animals can climb over Expensive

SOFTSCAPE

The softscape elements in your garden are the elements your chickens will most intimately relate to on a daily basis. Each item you choose to place in your garden, from the living soil to the smallest plants, will be a significant inclusion.

Softscape Materials

Healthy soil is the foundation to a healthy landscape. And so I tend to stress the importance of choosing your soil and mulch products carefully for the needs of your garden and for free-range chickens. As gardeners, we often have to add soil amendments to the existing soil to get a properly balanced, healthy growing medium for our plants. Chickens will do a great job of adding organic matter to soil over a period of time, especially if you compost their droppings and bedding. In nature, the process of building soil takes thousands of years, so as gardeners we need patience.

○ *Soil amendments.* To improve your soil, there are many types of soil amendments available to gardeners, whether it is homemade compost, a bagged product from a local garden center, or a product delivered in a bulk quantity from a landscape supply company. Companies have been known to sell inferior products with contaminants such as heavy metals, pesticide residues, and a weed seed bank that is a result of incomplete composting or poor raw materials. Do your research to learn about the product and the company, so you know exactly what you are getting. These problems can occur with homemade mulches and composts as well.

Depending on the soil type on your building site—compact urban fill, glacial till, or clay—you may need to amend it with specific elements to create a properly balanced growing medium. For example, my native soil is glacial till which is heavy in minerals with large-particle gravels, cobble, and some loam. In this type of soil, you can't dig a hole for a single plant without running into at least a dozen rocks about the size of your fist. Glacial till is lacking in organic matter, so I add compost, the most readily available soil amendment. I make compost on our site with the manure and bedding from our animals and with greens like leaves and clippings. After the organic material breaks down, it makes an

appropriate, rich, soil amendment. If you are not sure what your soil is lacking in nutrients, get a soil test before you purchase soil amendments.

○ *Mulches.* A big complaint of free-range chicken owners is that the birds fling mulch everywhere when they scratch in the soil, which can be messy and a chore to clean up. But mulch is one of the most important things for your garden, along with good design, healthy soil, and proper watering. In nature, plants provide their own mulch to the soil. Every fall, deciduous trees lose their leaves, which break down over the winter and are the perfect blend of nutrients for that plant, building the soil over time. Mulch is not only good for soil building and plant health, but your chickens will love eating the organisms living in it. Keeping your garden beds mulched will retain moisture in dry summer months, insulate and keep the soil warm in cold winter months, moderate extreme soil temperatures, prevent weed growth, feed the soil and plants, protect soil from erosion during heavy rain, protect soil from compaction, and encourage soil organisms that your chickens will eat.

A wide variety of mulch materials is available to the gardener. Certain mulches will perform better than others, depending on parent material.

There are many ways to make your own garden mulch. Plants make their own mulch, dropping and accumulating organic matter over time, like leaves or needles, if we don't disrupt the process as it is intended to work. The process takes years, and may not be the look gardeners want. Chickens will assist in this process by scratching the organic matter around and breaking it up, while leaving their manure behind.

In fall, we rake the leaves off the lawn and place them throughout the garden beds. They act as a natural insulation, and after they break down, they feed the soil. Be careful not to let leaf mulch rest up against the trunks of trees or against any woody stems of shrubs, and don't use diseased leaves that can infect the plants.

By using a compost system, you can create rich mulch from your garden waste, chicken manure and bedding, and kitchen waste. For most food scraps, we use worm bins, and the material becomes excellent mulch, topdressing, or fertilizer.

mulch materials

TYPE OF MULCH	PROS	CONS	AVAILABILITY
COMPOST (ORGANIC MATTER)	Rich nutrient with organic matter content Moisture retention	Weed seeds love to grow in it	Readily available, home-made or commercially produced
WOOD CHIPS	Long-lasting weed prevention	Can harm young, sensitive, herbaceous plantings	Can be free if you contact an arborist company or purchase from landscape suppliers
BEAUTY BARK	Readily available	Not always sustainably harvested and doesn't provide many nutritional benefits to the soil Can repel water	Readily available
PINE NEEDLES	Lightweight	Can be messy to work with and can look "unfinished" Can lower pH in soils with continuous use.	Free if you have a pine tree Available commercially only in certain regions
STRAW	Readily available Lightweight	Can contain weed seeds Breaks down quickly	Readily available
COCOA BEAN HULLS	Smells delicious and looks amazing	Toxic to some animals Expensive	Hard to find in large quantities Available only regionally
HAZELNUT SHELLS	Slugs and snails hesitate to cross over it Birds love the leftover nuts in the shells	Takes years to break down Expensive	Expensive to purchase Available regionally only; try contacting local nut orchards
ROCK	Chickens can't easily scratch in it Can create heat to warm soil	Can be difficult to weed Can create too much heat	May be sourced on site in existing soil Expensive to purchase

◦ *Edging.* Using an intentional edge to contain the mulch within beds is one way to prevent a messy mulch issue from chickens scratching. Edging comes in a variety of materials such as submerged flexible plastic or metal strips to lines of stones you find on site. I am not a fan of edging. As a designer, I think the line of the edge draws your attention to it and away from the plants or surrounding gardens. I also think edging is often simply ineffective: edging that is installed at or below the garden surface will not work with chickens, because they will just fling mulch right over it. But it only takes a few minutes to clean it up with a broom or rake, depending on the mulch. If you need to use edging, I would use natural materials, especially ones found onsite, like stones, large woody debris, or branches.

tip

Be sure to use edging that is at least a few inches above grade to capture the mulch as the chicken scratches nearby.

◦ *Synthetic materials.* In general, most synthetic and plastic materials don't belong in our gardens and are not needed. They will not break down as other natural materials will over time and could potentially be toxic. Because of toxicity concerns, I do not advocate using objects like old tires for raised beds, which would come in contact with edible plants or their roots. Product marketing is a powerful force in the gardening world, especially in retail nurseries or big box stores where they will try to convince you that you need their product as a gardening aid. Before you buy the "next best" product for your garden, consider what it is made of, whether the materials are biodegradable, and whether it could be potentially harmful to you or your chickens.

◦ *Landscape fabrics.* I dislike use of all synthetic materials in the garden, including landscape fabrics, especially as a weed barrier. If you are trying to prevent weeds long term by using it under mulch or gravels,

forget it. Landscape fabric will work for a season or two, but eventually weed roots will grow through, and when you try to pull those weeds a bubble of landscape fabric will come up through the mulch and look horrible. Then the landscape fabric is hard to fix without ripping up everything or tearing holes in it, leading to a complete waste of time, money, and effort. I also do not use landscape fabric as a weed barrier because it doesn't allow soil organisms to move up through the different layers of soil into the mulch, to help break it down, which is essential for healthy soil. When chickens are scratching, they will lift it and their manure will become embedded in the fabric, making it a potential breeding ground for bad bacteria.

There are, however, some beneficial uses for fabrics in gardens. Dark-colored fabric often made of polypropylene can be used in the early spring on top of your vegetable beds as a way to warm the soil. It will allow air and water to flow through, but the dark color will retain heat from the sun in those days when we need it most in preparing our vegetable garden for spring. When you are ready to seed or plant, it can be folded and put away for the next year's use. Some fabrics and plastic sheeting are used as mulch in large-production vegetable and berry beds as a way to retain heat and keep weeds down. Light-colored fabrics or horticultural fleeces like Remay cloth are used to keep crops protected in cold weather and when pests are a potential problem. These frost blankets can be useful in an unexpected early or hard freeze and can also be reused.

tip

As an alternative to using fabrics as a weed barrier, I recommend sheet mulching with a material like newspaper or cardboard under mulch, which helps inhibit weed growth until it decomposes naturally.

OPPOSITE: A daisy grows in an eco-lawn.

Lawn

Lawn is easily the most controversial element in the landscape and garden industry. Turf is heavily dependent on natural resources like fresh water and fossil fuels to keep it at its perfect 2- to 3-inch height year round. In the quest for the perfect lawn, homeowners annually dump millions of pounds of polluting fertilizers and pesticides to keep the grass weed free and perfectly green. While lawn is often an unnecessary element in many gardens, it is actually an important design element to consider for a garden with free-range chickens.

Chickens and organically maintained lawns will have a good symbiotic relationship if they are managed well. Chickens graze on the grass blades as a choice food source for greens, which can also be collected and dried for food. Chickens will keep the lawn mowed, so to speak, because they will make the lawn a frequent stop in their daily foraging. Chickens will search out and eat insects in lawns, and leave behind their manure, which is an excellent fertilizer for lawns. The height of your lawn is critical: the blades need to be long enough to photosynthesize in order to be healthy and grow—3 inches is ideal; and chickens prefer a certain height of grass blade—5 to 6 inches is too tall for them to graze, and letting them graze below 2 inches can damage the grass. So for chicken foraging, it is best to keep your lawn about 3 to 4 inches, which may be a bit longer than you are used to cutting it, but the turf will be healthier and withstand the chicken's grazing better.

go greener with eco-turf

Consider growing an eco-friendly lawn, instead of the conventional, perfect lawn, to keep your hens and soil their healthiest. Ecological seed mixes, which you can get at your local retail nursery or feed store, contain a variety of plant types, many of which might be considered weeds to a gardener who prefers a perfect monocrop of grass blades, but are an excellent, diverse source of greens for chickens to forage. An eco-lawn requires less maintenance such as mowing, and requires fewer resources like water and fertilizers. Clover, one of the most important plants found in eco-seed mixes, fixes nitrogen naturally in the soil and eliminates the need for fertilizers. For my own lawn, and I use that term loosely—I let just about anything grow. If it is green and not a noxious weed, it is welcome. I do not use pesticides and I don't mind moss, clover, and other broadleaf evergreens, because they are less work. Besides, lawn isn't my focus: it's the plants and the structure.

clover

Clover (*Trifolium* species) has been called one of the best all-around feeds for animals because of the high nutrient value. And it also is a legume, so it fixes nitrogen in the soil naturally. Clover is also an important nectar source for bees and a host of other pollinating insects. There are several types of clovers, including white Dutch clover (*Trifolium repens*), which is low growing and requires less mowing than a conventional lawn, making it a great addition to eco-turf seed mixes. Clover is also good to use in seed mixtures with grain as a companion plant because it loosens subsoils and makes more nutrients available. If pastures are being shared with large livestock, there must be diversity in what is grown, since a field of pure clover can make a cow, horse, sheep, or other animal bloat.

OPPOSITE: White Dutch clover.

plants with purpose

Plants are complex living organisms that breathe life into our planet. We eat plants, and we use them for shelter, construction, textiles, and medicine. Without plants we wouldn't exist. My passion for plants goes well beyond aesthetics and reaches into the countless functions plants can have. Through time, human civilization has discovered many ways in which plants can be of benefit to us. Learning about the traditional roles of plants can teach us a lot about how we can use plants today.

angela's garden

IN THE HEART OF TOWN, there is an appealing little garden that's hidden behind a formal green wall bordering the street. In the hedge, there's a door so you can pass through, and behind that door lives a small flock of free-ranging Silky chickens, three hens and one rooster named Buttercup. Angela and Travis, a young couple, allow their birds to do as they please on their small urban lot (less than 1/10 acre) that has it all—perennial beds, eco-lawn, kitchen garden, entertaining patio space, and a charming little greenhouse built from salvaged window panes.

The flock spends most of their day hanging out in the garden, foraging and taking shelter under shrubs while Angela gardens in her rubber boots. She has a special little black hen named Licorice that behaves much differently from the other chickens, following Angela around like a puppy. Occasionally she even gets to go inside the house.

The garden is constantly changing, with new plantings and art. The small chicken coop is positioned in an area that is visible from inside the house so it is easy to keep an eye on the flock. Atop the coop is a "green roof" covered with selected plants like sedums and hens and chicks (*Sempervivum tectorum*), and a nearby bog garden has carnivorous plants that eat any flies coming from the coop area.

OPPOSITE: Silky rooster Buttercup struts in front of a perennial border.
NEXT PAGE, CLOCKWISE: Licorice weeds the sandset-tumbled bluestone patio. A Silky's foot has five toes and a feathered leg. A formal osmanthus hedge makes a beautiful and effective green wall for a chicken garden.

PLANTS HAVE DIFFERENT AESTHETIC QUALITIES at different times of year, whether it is spring bloom or fall color. When selecting plants for our chicken garden, we also need to consider whether they have these functions and characteristics: soil building (nitrogen fixing and dynamic accumulators), shade and shelter, food source, medicine, insect repellant, cleaning and cosmetic properties, windbreak and wildfire protection, wildlife habitat, nectar source for pollinators, and fuel source.

PLANTS FOR CHICKEN GARDENS

As gardeners, we are usually limited to a certain amount of space and can't grow everything we would like to grow. In garden design, not only do we consider the basics for choosing the right plant for the right place (mature size, exposure, soil needs), but as gardeners with chickens, we need to think about how each plant's special characteristics will function for us. When it comes to designing for chickens in your landscape, it is best to choose plants that provide as many functions as possible. Plants that produce food for us can also provide food for our chickens. Plants that provide shelter for us also provide shelter for our flock.

weedy precautions

Invasive species vary from region to region. What is a highly prized plant in one area may be a menace in another. Some plants considered weeds by gardeners may just be plants you don't want in certain locations. But noxious weeds can be destructive and are difficult to control, often spreading into the ecosystem or causing financial hardship in the economy. Noxious weeds are listed on federal, state, county, or other local websites. Before buying and planting anything that is new to you, be sure to check whether it will wreak havoc in your local ecosystem.

Soil-Building Plants

Gardeners are always looking for ways to improve their soils. One great way to build the soil is to use plants that make specific nutrients available to feed our plants and in turn to feed us and our chickens. In a chicken garden, it is important for our soils to be as healthy and rich as possible, so that we can grow high-demand plants for our food and for our foraging animals. Many good chicken plants are either nitrogen fixers or dynamic accumulators.

○ *Nitrogen-fixing plants.* Nitrogen fixers are plants that make rich nitrogen available to the soil through a beneficial relationship with bacteria that lives in nodules on the roots of those specific plants. Nitrogen fixers are often pioneer species and many are from the legume family. There are many trees, shrubs, perennials, and annuals that are nitrogen fixers, such as alder, bayberry, clover, and peas.

○ *Dynamic accumulators.* These plants are similar to nitrogen fixers in that they help make specific nutrients available. Their roots drill deep into the soil and pull specific nutrients into their leaves and roots, which can then be used for mulch or to create compost or fertilizer. There are many dynamic accumulator plants, including borage, comfrey, marigold, and yarrow.

Habitat

For free-ranging chickens, our primary aim in designing the garden is to provide habitat. The best landscape design for wildlife was created by Mother Nature, so it is good to mimic what has worked in naturally created ecosystems. While native habitat for wild chickens was in thick forests of subtropical regions, we can create the basic elements in all kinds of climates and in our own backyards.

OPPOSITE, TOP TO BOTTOM: Apples ripen in a food forest at the garden. The black currant shrub is armed with impressive thorns.

The four main puzzle pieces chickens need for a safe and nurturing environment, whether they are wild or domesticated, are:

o *Shelter*—from predators and extreme weather
o *Food*—plants they can eat (seeds, fruit, nuts, greens)
o *Water*—fresh water at all times
o *Nesting sites*—safe for raising their young

Even if it is not our intention to raise chicks, creating nesting sites (usually in the coop) will give the hens a designated space to lay their eggs, and they need to have access to that space and feel safe in it. If their nest boxes get too dirty or crowded, they may look elsewhere, so be sure to keep the space clean and big enough for your hens.

Shelter

Chickens need shelter, from the weather and from predators. In the wild, chickens rely on plants to hide in and under, so if your chickens free range, they'll need someplace to take cover at all times. Trees can allow refuge from ground predators, while shrubs provide cover as protection from overhead predators. Many shrubs make perfect shelter, and they can provide food for us, for other animals, and for pollinators, including:

o *Evergreens.* Plants that keep their leaves year round are ideal in areas with longer winters when all of the deciduous plants have lost their leaves and provide less protection. Free-range chickens will run quickly to hide when they hear loud overhead noises and if they are in evergreen shrubbery, it's less likely they will be seen and attacked.

o *Thorny thickets.* Thorny thickets are one of the best shelters we can provide for free-range chickens. Large predators, especially airborne ones, will avoid getting tangled up in a mess of thorns, but some small ground predators might not be completely deterred. Plants that have spiny leaves are also a good deterrent. Parents of young children often prefer that these plants are not in their gardens, but kids learn quickly to avoid them, and these plants could mean safety for your flock.

OPPOSITE: A chick peeks out of the feathers of its protective mother hen.

shelter plants for chicken gardens

○ **CANE BERRIES:** Blackberry, raspberry, salmonberry (*Rubus species*). This genus contains many deciduous shrubs or cane plants with upright branches, berries, and thorns. Cultural conditions vary with each variety. Zones 4–8.

○ **CHINESE HOLLY** (*Ilex cornuta*). This evergreen shrub or small tree with glossy green spiny leaves is not as hardy as English holly (*Ilex aquifolium*) but is a good alternative to boxwood. Zones 7–9.

○ **DARWIN BARBERRY** (*Berberis darwinii*). This hardy evergreen grows to be almost 10 feet tall, has dense yellow-orange flowers in spring, small hollylike spiny leaves, and dark purple edible fruit. Zones 5–9.

○ **ENGLISH HOLLY** (*Ilex aquifolium*). This evergreen tree or shrub, which comes in several varieties, will take a shrub form if pruned in early years, is slow growing, and has attractive winter fruit. Zones 3–7.

○ **GOOSEBERRY** (*Ribes hirtellum*). A deciduous shrub that is armed with thorny stems, gooseberry produces a tart edible grapelike berry. There are many gooseberry cultivars, and the shrub is prone to a few pests. Zones 3–8.

○ **HAWTHORN** (*Crataegus species*). The mostly small trees or shrubs in this large genus are sometimes called thornapple. They have thorns on their stems, white flowers, and small edible fruit that has been used for medicinal purposes. Zones 4–9.

○ **JAPANESE BARBERRY** (*Berberis thunbergii*). This deciduous thorny shrub has many popular cultivars ranging in size and attributes. It has yellow flowers in late spring and produces small edible red fruit that usually persists into winter before birds get all of it. Zones 5–8.

○ **JUNIPER** (*Juniperus species*). These evergreens, which have needlelike or scalelike leaves, will tolerate a wide range of climates and soil conditions. There are many varieties available. Zones 3–9.

○ **LONGLEAF MAHONIA** (*Mahonia nervosa*). This evergreen shrub with spiny, toothed leaflets spreads underground, creating a thick groundcover. Zones 6–8. *Mahonia repans* is a similar species.

○ **MAGELLAN BARBERRY** (*Berberis buxifolia*). This evergreen shrub has leathery, spiny leaves. It has upright growth, yellow flowers in mid to late spring, and edible fruit that is said to be the best tasting of all barberries. Zones 5–9.

○ **OREGON GRAPE** (*Mahonia aquifolium*). Oregon grape is an evergreen shrub with erect habit, compound leaves with very spiny, toothed leaflets, yellow flowers in spring, and purple, edible fruit. Zones 5–9.

○ **RUGOSA ROSE** (*Rosa rugosa*). Compared to most ornamental roses, this deciduous or evergreen shrub is pest resistant and dependable. Its very prickly stems and large, useful rose hips are great for a chicken garden. Zones 3–9.

○ **SCARLET FIRETHORN** (*Pyracantha coccinea*). This evergreen shrub has thorns that leave a sting. The dense irregular growth habit is often sheared or espaliered. It flowers in early summer, and then has showy red berries. Zones 6–9.

○ **THORNY ELEAGNUS** (*Eleagnus pungens*). This evergreen shrub has dark green, leathery leaves, upright habit, dense growth, and spiny branches. It has silver-white flowers in the autumn. Zones 7–9.

○ **WARTY BARBERRY** (*Berberis verruculosa*). This evergreen shrub has yellow flowers, blue-black fruit, dark green leaves, and sharp spines. Zones 6–9.

○ **WINTERGREEN BARBERRY** (*Berberis julianae*). This evergreen or deciduous plant can form a very dense compact shrub, with spiny thorns under the leaves that are one of the thorniest. Zones 5–9.

OPPOSITE: Hens love to snack on sunflowers.

FOOD AND FORAGE

In our backyards, we don't have endless natural forest habitat for our chickens, but we can try to mimic that habitat by growing a food forest garden of specific plants that supplement their feed. While free-range chickens can forage much of their food, you will most likely not be able to achieve complete independence from providing feed. The nutritional needs of chickens are protein, carbohydrates, fat, minerals, vitamins, and enzymes, as well as grit to help them break down their food. By letting your chickens free range, you allow them to choose and eat what they need. Not all chickens make the best choices, but if given the chance, most will learn to eat only what they need. Chickens thrive on variety.

Grains

One chicken can consume roughly one bushel of grain per year. Grains can be a valuable crop to a homesteader. Having a meaningful amount of grain to harvest means that you need a specific amount of land for growing the grain and know how to harvest and store the grain. For example, you can plant 6 pounds of wheat seed in an area that is 20 by 50 feet in your garden and harvest close to 50 pounds of grain. That 1000 square feet of garden space might be better used for growing more diverse crops with higher yields. Each grain is different, and has different needs and harvesting requirements. While grains are not the most sustainable crop—being monocrops with relatively high input requirements—they can still be a very good food source for us and for our animals. Grains also have another wonderful use: their leaves and stalks can be used for straw bedding and for composting or mulch. This benefit alone makes the crop valuable to the chicken gardener, who might otherwise spend a lot of money bringing that material in from an outside source.

Chickens also don't necessarily need the same processing that we need to harvest and eat grains ourselves. Their beak is designed for eating seeds, and they can scratch through the roughage to find the seeds. There are many grains chickens like, and several ways to feed chickens grain, whether it be the seed—whole, processed, soaked in water, cooked, grown for greens to forage—or germinated to sprouts. The list shows grains in the order they are most commonly grown for chickens.

tip

Wheatgrass, or young wheat plants about 4 inches tall, is a popular juicing ingredient known for its nutritional values. It can be grown in flats to be fed to chickens, which works well if you keep your chickens in a coop and run without free-range access. This micro pasture method works really well: once the chickens have grazed the flat of wheatgrass, replace the flat with one that has had a chance to regrow.

you are what you eat

When it comes to our food—eggs and meat especially—it is important to consider the animal that is creating our food and what they have been eating. After all, we are eating what they have eaten, just after it was processed through their bodies. If our backyard chickens are raised to eat only commercially grown feed, that could be the same feed that factory farm chickens eat. Chickens all have unique preferences for what they eat. Given lots of choices, they will tend to eat what they want and skip over food that does not appeal to them. When chickens are foraging, I've seen one chicken love a plant that another chicken completely avoids, and a plant that is devoured in one year will be avoided in the next year. What chickens eat will surprise you, and many factors can affect chickens' foraging choices:

○ Are they hungry? How often are they fed and how much? Do you leave out full feeders or do you make them search for their own food?

○ What variety and quantity of food sources are available to them at any given time?

○ What is their experience at foraging? Older birds usually know where the "good" food is. Inexperienced birds tend to eat anything just to try it.

There is no guarantee that your chickens won't eat your prized plants unless you protect them, while avoiding the plants and food that you provide for them. But the more variety of plants for the chickens to forage from, the less likely you will have chickens eating plants they shouldn't be eating.

OPPOSITE: A hen forages in the soil along a path.

grains for chickens

○ **CORN** (*Zea mays*). This crop is a great beginner grain to plant, since the seeds germinate readily and it will grow quickly in the right conditions. There are many types of corn to grow, all having different maturing times and uses or flavors. Corn needs full sun, heat, plenty of water, and is a nitrogen hog, so be sure that the corn plants are well fertilized or have nitrogen-fixing plants nearby. Corn can be processed and stored in a number of ways, but the easiest is to store mature harvested cobs whole and feed them to the chickens broken into chunks.

○ **WHEAT** (*Triticum aestivum*). Wheat is another easy crop to grow, but processing wheat can a bigger task. There are many different kinds of winter and spring varieties of wheat. Sowing the seed is similar to starting a new lawn, but should be worked in 1 or 2 inches. Wheat can provide good forage for chickens, and good cover toward the end of the growing season. Whole wheat seeds can be eaten by grown chickens, but baby chicks will need them ground. Compared to corn, wheat takes longer to mature, it has more protein but fewer carbohydrates, and it makes a good scratch mixed with corn.

○ **OATS** (*Avena sativa*). This grain is easy to grow but one of the hardest to hull. Known for the impressive protein content, oats are widely grown for human and animal food. By simply soaking oats in water, they become more digestible to chickens. Crimped oats are also more digestible, and many chicken keepers often make oatmeal for their chickens. Other good oat species for chickens include naked oats (*Avena nuda*) and red oats (*Avena byzantine*).

○ **BARLEY** (*Hordeum vulgare*). This versatile grain can be grown in just about any region, warm or cool. It contains three times more protein than corn and is sometimes a replacement for corn in livestock feed. Chickens can be picky eaters when it comes to raw barley if there are other options available, but they can eat the barley whole without processing. They prefer the barley to be sprouted, but it needs to be "after-ripened," or go through a process after harvest before they can be germinated, which can take up to 6 weeks. Like wheatgrass, barley grass has become a popular juicing ingredient.

○ **MILLET** (*Setaria italica*). Of the many millet types, foxtail millet is most commonly used as chicken forage and is found in a lot of bird food mixes. Like corn, millet requires a lot of nitrogen from the soil and prefers hot summers. Chickens will eat the seeds, can eat the grass too, and will also find shelter in it as it grows.

○ **WINTER RYE** (*Secale cereale*). Winter rye is a good grass crop for chickens to graze, but is not typically used commercially for grain scratch or feed. It is in many cover crop mixes, as well as green manure mixes, which when grown are turned back into the soil to add back certain nutrients. Known for its hardiness, winter rye can germinate in temperatures as low as the 40s, will grow in poor soils, and is a good crop in pastures.

○ **BUCKWHEAT** (*Fagopyrum esculentum*). A good green manure or cover crop, this grain does best in cooler climates and will tolerate a lower pH in soils. It grows quickly and works well for smothering weeds. It has high amounts of a protein called lysine, which is an essential amino acid. Chickens can eat buckwheat, but reports of photosensitivity in larger animals keep some chicken owners from feeding it to their birds without cooking it first.

sunflower

Sunflowers (*Helianthus annuus*) are a garden classic that produce tasty, nutritious seeds for you and your flock. With many varieties to choose from, this annual plant is easy to grow in just about any garden. Be sure to plant them in an area with full sun and well-drained soil. And remember that many varieties will grow very tall, creating shade to the north of them, so plant them in the northernmost part of your garden or where you need to create shade. Chickens love to eat sunflowers straight from the heads. If you want to save them for your family, when the leaves turn brown simply cut the head with a few inches of stem so you can hang them in a dry place like the garage, much like you would for garlic or onions. You can leave them on the stem in the garden, but you may need to put netting around the heads as protection since wild birds and squirrels also love sunflower seeds. Oil can also be rendered from the seeds, and the stalks and leaves can be used as chicken bedding or composted into mulch.

OPPOSITE: Sunflowers produce large amounts of nutritious seeds.

Legumes

Legumes make a good companion plant to broadcast with grain seeds or to plant in rotation because of their ability to fix nitrogen naturally. Many grains deplete the soil of nutrients, so planting legumes like clovers, vetch, or soybeans will help prevent that by rebuilding the soil. Alfalfa is another good perennial legume to feed chickens. Some chicken owners prefer to feed alfalfa pellets that have been moistened with water.

Other seed-producing plants that can be grown for chicken feed are sunflower, amaranth, chickpea, and sorghum. They can be sown in rotation in a paddock system. It has been said that amaranths seed causes illness in poultry, but recent reports say that would be true only in large quantities, and that there are many positive benefits from adding amaranth seed to chickens' diet.

Trees

Just about any fruit-bearing tree is a good addition to a chicken garden. The flock will help clean up any fallen fruit, keeping potential pests at bay. When chickens scratch they rarely damage trees once they are established. When you are selecting a fruit tree, keep in mind that for every fruit type, there are many cultivars that may perform differently in your area regarding frost or pest and disease control. Varieties also have different mature sizes and fruit that is unique to that variety. Fruiting trees may or may not require a pollinator tree in order to produce fruit, so be sure to check into that before purchasing.

Trees that provide food for chickens in the form of seeds or nuts can also be useful as windbreaks, shelter, or valuable lumber used for fuel or construction. The honey locust tree (*Gleditsia triacanthos*) is a nitrogen fixer which produces seed pods and leaves that can be eaten by chickens as well as other animals. The tree also provides good bee forage and shelter as it is covered in thorns when young. Thornless varieties are available.

Nuts from various trees can be stored and fed to chickens in the winter months in moderation. Nut trees are generally large, slow growing, and might require two trees for cross-pollination. You may find competition from squirrels during harvest time. Many shelled nuts can be fed directly to chickens, such as peanuts, pistachios, and cashews, but do not feed chickens salted or flavored nuts. Some contain tannins, so they need to be processed before feeding them to birds.

OPPOSITE: I mix edibles with ornamentals in the garden. Zucchini adds a nice bold texture.

fruit trees for chicken gardens

○ **APPLE** (*Malus pumila*). This popular tree with delicious and nutritious fruit comes in many varieties. Hardy to Zone 4.

○ **APRICOT** (*Prunus armeniaca*). This small tree has a beautiful early spring bloom and sweet, tender fruit. Hardy to Zone 4.

○ **CHERRY** (*Prunus species*). There are many varieties to choose from, from sweet to tart. Hardy to Zone 5.

○ **CORNELIAN CHERRY** (*Cornus mas*). This fruit tree is highly ornamental and blooms early in the spring. Its fruit can taste like plums or cherries, depending on the variety. Hardy to Zone 4.

○ **CRABAPPLE** (*Malus coronaria*). This dwarf tree is known for its striking flowers in spring and small red fruit. Some varieties are more pest and disease resistant than others. Hardy to Zone 4.

○ **FIG** (*Ficus carica*). This tree offers tropical-looking foliage and delicious fruit, and is deer resistant. Hardy to Zone 7.

○ **MEDLAR** (*Mespilus germanica*). This tree is self fertile, disease resistant, ripens late, and produces a fruit that tastes like applesauce. Hardy to Zone 4.

○ **MOUNTAIN ASH** (*Sorbus aucuparia*). This tree has small clusters of nutrient-rich fruit that ripens in the fall. Hardy to Zone 3.

○ **MULBERRY** (*Morus species*). This tree produces blackberrylike fruit. Hardy to Zone 4.

○ **PAWPAW** (*Asimina triloba*). This tree produces a banana-flavored fruit and is pest and disease resistant. Hardy to Zone 5.

○ **PEACH** (*Prunus persica*). This tree, and nectarine (*Prunus persica* var. *nectarina*), blooms early and produces delicious fruit. In areas prone to wet springs, choose a disease-resistant variety. Hardy to Zone 5.

○ **PEAR** (*Pyrus communis*, European, or *Pyrus serotina*, Asian). This fruit tree comes in many varieties and is just as ornamental and beautiful as it is functional for food. Hardy to Zone 4.

○ **PERSIMMON** (*Diospyros species*). This small tree has tropical looking foliage and tasty red fruit. Hardy to Zone 7.

○ **PLUM** (*Prunus species*). This common tree produces abundant sweet fruit. Some varieties are self-fertile while others require a pollinator. Hardy to Zone 4.

○ **QUINCE** (*Chaenomeles species*). Quince shrubs are known for their large flowers and nutritious fruit. Hardy to Zone 4.

nut trees for chicken gardens

○ **ALMOND** (*Prunus dulcis*). A smaller tree producing a nutlike seed, almond works well in an urban garden with a height maintained around 15 to 20 feet. Almonds can be preserved for 1 to 3 years depending on how they are stored. They can be mashed for chickens to eat with other grains. Bitter almonds (var. *amara*) can be toxic to birds in large quantities. Zones 4–9, depending on variety.

○ **CHESTNUT** (*Castanea* species). This genus contains several species of large trees that can grow over 100 feet tall. Chestnuts can be eaten and prepared in many ways, but storage is limited to 6 months unless they are chilled. Zones 5–9.

○ **FILBERT** (*Corylus* species). Filbert, or hazelnut, is a nut-producing tree that stays within a reasonable size for urban gardens. It prefers a wet, cool climate. Filberts can be stored for up to a year. Zones 3–9.

○ **OAK** (*Quercus* species). This genus includes many species, which range in size and cultural requirements. Generally, oak is large, slow growing, and produces acorns with high levels of tannins. The nut doesn't store for long periods of time, but can be dried and should be cooked and mashed before feeding it to poultry. Oak leaves can be shredded and used for mulch or bedding. Zones 3–9.

○ **WALNUT** (*Juglans* species). Walnut is a valuable timber tree, and produces a nut that can be stored for long periods of time, generally up to 1 year. The tree is allelopathic, which prevents some plants from growing under them, but in pastures the tree provides good shelter and habitat. Zones 4–9.

Shrubs

Shrubs that produce fruit or seeds are good choices for the chicken garden because the birds will clean up the fallen food source as it ripens. Just like other fruiting plants, be sure to select varieties that are right for your region and site.

shrubs for chicken gardens

○ **ARONIA** (*Aronia melanocarpa*). This white-flowering shrub bears nutritious small black fruit and has rich fall color. Hardy to Zone 3.

○ **BLUEBERRY** (*Vaccinium corymbosum*). This popular shrub provides both delicious berries and year-round interest in the garden, with many varieties to choose from. Hardy in Zones 3–11, depending on variety.

○ **CURRANT, GOOSEBERRY, JOSTABERRY, AND WORCESTERBERRY** (*Ribes* species). This genus has many species and varieties that flower in early spring and bear different colored and different tasting fruit. Hardy to Zone 3.

○ **ELDERBERRY** (*Sambucus* species). This medium-sized shrub is a favorite of the birds. It bears small nutritious berries that range from black to blue to red in color, depending on species. Hardy to Zone 4.

○ **GOJI BERRY, OR WOLFBERRY** (*Lycium barbarum*). This shrub has light purple flowers and bears small, red fruit rich in antioxidants. Hardy to Zone 5.

○ **HAWTHORN** (*Crataegus* species). This genus of mostly small trees or shrubs are sometimes also called thorn-apple because they have thorns on the stems. Hawthorn has white flowers and small edible fruit. Zones 4–9.

○ **HONEYBERRY** (*Lonicera caerulea* var. *edulis*). This small shrub produces a blueberrylike fruit that ripens very early. Hardy to Zone 3.

○ **RASPBERRY** (*Rubus idaeus*). This delicious cane fruit comes in several varieties with different ripening times and different colored berries. Hardy to Zone 3.

○ **RUGOSA ROSE** (*Rosa rugosa*). Compared to most ornamental roses, this shrub is pest resistant and dependable. Its prickly stems and useful rose hips are excellent for a chicken garden. Zones 3–9

○ **RUSSIAN OLIVE** (*Elaeagnus* species). This large adaptable shrub has aromatic yellow flowers followed by silvery fruit that birds love. Hardy to Zone 2.

OPPOSITE: A hen eats blueberries off the shrub.

◦ **SASKATOON BERRY, OR SERVICEBERRY** (*Amelanchier* species). This large shrub produces white spring flowers, edible fruit, and fall color, making it a great ornamental. Hardy to Zone 4.

◦ **SEA BUCKTHORN** (*Hippophae rhamnoides*). This shrub can grow in poor soils and produces clusters of small orange fruits that can be squeezed to make a substitute for orange juice. Hardy to Zone 3.

◦ **SIBERIAN PEA SHRUB** (*Caragana arborescens*). This deciduous shrub is commonly used in chicken yards as a dependable cold- and drought-tolerant, multifunction plant. It fixes nitrogen while providing thorny shelter and edible seedpods. It grows up to 15 feet in a well-drained sunny location, and has small yellow flowers in spring that are a favorite of the bees. Zones 2–9.

BELOW: Elderberries are a chicken's favorite.

Vegetables

Just about any food we grow in our vegetable garden can be fed to our chickens. Some vegetables like beets and potatoes need to be cooked or cut into small pieces for the chickens to enjoy. Other crops can be eaten by the chickens with our help by simply cutting them open to expose the soft flesh, such as pumpkins, squashes, and melons. Free-range chickens can be beneficial in the vegetable garden by weeding, and they eat damaging pests such as cabbage worm.

Greens and seeds are an important part of a chicken's diet. Many greens can be grown in an eco-friendly lawn mix, or within an area where you don't mind a bunch of greens going to seed and taking over. Greens can be harvested for the chickens or they can have direct access to the area where they are grown. Many plants that provide good green forage for chickens are considered weeds by conventional gardeners and are not sold in nurseries but can be purchased through catalogs. But most are common wild plants and can be found in gardens and easily propagated. You may already have many of these plants in your garden and consider them weeds, but you can now think of them as useful chicken plants.

> **tip**
>
> *When vegetable crops are ripe or close to ripening, fence those areas off. You can also try planting off-colored crops such as yellow tomato varieties that the chickens seem to avoid.*

greens for chicken gardens

◦ **BIRD'S FOOT BROADLEAF TREFOIL** (*Lotus corniculatus*). This legume, which doesn't cause bloat, is used in pastures and grass mixes.

◦ **CHARD** (*Beta vulgaris*). Also known as Swiss chard or beet chard, this vegetable can be grown as forage or fed to the chickens.

◦ **CHICKWEED** (*Stellaria* species). Chickweed is a common weed that is a favorite chicken forage plant.

◦ **CHICORY** (*Cichorium intybus*). An herbaceous perennial, chicory can be grown as a forage crop.

○ **CLOVER** (*Trifolium* species). Clover, including ladino, red, strawberry, and white Dutch clover, is a good forage plant and a valuable nitrogen fixer.

○ **COWPEA** (*Vigna unguiculata*). Also called southern pea, blackeyed pea, and crowder pea, this green manure crop or cover crop attracts beneficial insects, suppresses weed growth, and produces an edible pea that chickens can eat.

○ **CORN SALAD** (*Valerianella locusta*). Also called lamb's lettuce, nut lettuce, and rapunzel, this low-growing winter green is a good chicken forage plant.

○ **DANDELION** (*Taraxacum officinale*). One of the most common weeds, dandelion is also good chicken forage.

○ **DOCK** (*Rumex* species). Also called sorrel, dock is a common weed that can be used as forage for poultry.

○ **FLAX** (*Linum usitatissimum*). The plant bears a nutritious seed, and can be grown and harvested much like grain.

○ **LAMBSQUARTERS** (*Chenopodium album*). Also called goosefoot, this common weed is a tasty, nutritious food plant.

○ **LETTUCE** (*Lactuca sativa*). One of the easiest edible annuals to grow, lettuce can be fed straight to your chickens or can be added to a forage seed mix.

○ **MUSTARDS** (*Brassica* species). This genus of vegetables, including radish, mustard, spinach, broccoli, cabbage, kale, cauliflower, brussels sprouts, turnip, bok choy, and kohlrabi, will produce lots of food for you and your chickens.

○ **PLANTAIN** (*Plantago* species). A common weed, plantain is good chicken forage.

○ **PURSLANE** (*Portulaca oleracea*). Also known as pigweed or little hogweed, this little weed when foraged by chickens is reported to raise the omega 3 level in eggs.

○ **PIGEON PEA** (*Cajanus cajan*). This legume can be used as both a cover crop and a food crop for you and your birds.

○ **SESAME** (*Sesamum* species). This genus of annual herbs, which grows in warm regions, produces a seed that we use in cooking and is nutritious for poultry.

○ **SHEPHERD'S PURSE** (*Capsella bursa-pastoris*) has been said to prevent prolapse (a condition in a laying hen in which her oviduct doesn't retract after laying an egg) in chickens that free range and eat this plant.

○ **SORGHUM** (*Sorghum* species). This grass produces a seed that can be fed to poultry

Vines

Vines can be used as both shelter and food if the supporting structure is designed well. Chickens usually don't bother with plants that are a few feet off the ground, and will glean any fruit that falls onto the ground. Take care to protect the vine at its base so the roots don't become scratched up by the chickens.

vines for chicken gardens

○ **AKEBIA** (*Akebia quinata*). This semievergreen vine produces beautiful flowers followed by edible fruit. It needs a pollinator. Zones 4–9.

○ **CHAYOTE** (*Sechium edule*). Also known as choko, chowchow, or pear squash, this fruit-bearing vine grows in warmer climates. Zones 9–11.

○ **KIWI** (*Actinidia* species). There are many woody varieties of this vine that bears tasty fruit, including the arguta hardy species that can be eaten whole. Zones 5–9, depending on variety.

○ **GRAPE** (*Vitis* species). A popular fruiting vine, grape comes in many varieties, table grapes to wine grapes. Zones 5–9, depending on variety.

○ **MAGNOLIA VINE** (*Schisandra* species). This vine does well in shady, moist areas and produces an edible berry. Zones 4–8

○ **PEA** (*Pisum sativum*). Peas are another great choice for a vertical edible. Sweet pea seeds are toxic, so do not plant them over a chicken run with no other vegetation.

○ **PASSIONFLOWER** (*Passiflora* species). This tropical vine produces a stunning flower that is followed by an edible fruit. Zones 9–11; *P. incarnata* (maypop) is the most hardy.

○ **SQUASH** (*Cucurbita* species). Squash can be grown upward as a climbing vine. Some heavy fruit may need to be supported as they ripen, depending on variety. Annual.

○ **TOMATO** (*Solanum lycopersicum*). Tomato plants are often trained upward as vines for maximum production, and do best like that with chickens around, making the ripening fruit out of reach. Don't use tomatoes as a vine on a run fence without other foliage, since the leaves are toxic and chickens will be tempted to eat them if no other plants are nearby. Annual.

MEDICINAL PLANTS

For centuries, humans have relied on the medicinal characteristics of plants to remedy various ailments. Many plants known for their medicinal properties can also be beneficial for poultry ailments. Many of these plants are considered common weeds that we remove from our gardens. Keep in mind that each plant has different properties and may need to be prepared differently to extract the medicinal elements of the plant.

○ **CATNIP OR CATMINT** (*Nepeta cataria*). A perennial in the mint family, catnip is known for attracting pollinators, birds, and cats but repelling rodents. Its aromatic foliage has insect repellant properties. It can be used to prevent lice, fleas, and ticks on chickens. Many horticultural varieties of *Nepeta* species on the market have beautiful blue flowers and will bloom throughout the summer. 'Walkers Low' and 'Six Hills Giant' are two common varieties available in nurseries. Hardy to Zone 3.

○ **CHICKWEED** (*Stellaria media*). A common garden weed, chickweed has a long history of herbal use for many health issues, including improving circulation and reducing inflammation. The plant is a dynamic accumulator of potassium, phosphorus, and manganese, and chickens love to eat it, hence the name chickweed.

○ **COMFREY** (*Symphytum officinale*). This perennial herb is often used in permaculture design. It can grow quickly to 4 feet tall in many soil types and exposures. The large, bold leaves are edible for chickens, and the fibrous, deep-growing roots draw nutrients from the subsoil, making it an excellent dynamic accumulator. Comfrey creates a large amount of biomass every year, making it nutrient-rich mulch source and fodder. Once established, the plant can handle being cut back several times a year. Comfrey has a wide range of medicinal uses such as curing digestive disorders. But it is most known for containing allantoin, a cell-proliferating substance that speeds healing of broken bones, sprains, and cuts. Allantoin is synthetically re-created in the pharma-ceutical industry and used in healing creams. A homemade poultice can be made by pulverizing the plant—leaves, stems, and root—in a blender. Hardy to Zone 5.

○ **FEVERFEW** (*Tanacetum parthenium*). A short-lived perennial herb commonly known as a weed, feverfew will self-seed in our gardens. Its small daisylike flowers can be dried and used as an insecticide, having properties similar to pyrethrum, which is a plant extract with insecticidal properties traditionally used to help control insect problems. It has a good reputation as a medicinal plant. As the name may suggest, it is said to be helpful in curing fevers and headaches, and that the leaves can be placed directly on the gums for toothache. Hardy to Zone 6.

○ **GARLIC** (*Allium sativum*). Garlic is an easy-to-grow bulb that is used widely for culinary purposes and is known for its pungent flavor and aroma. The health benefits of garlic are considered to be many: it has antibacterial, antiviral, and antifungal properties. It is also known to stimulate the digestive organs, and to fight the common cold and cough. If small crushed pieces of garlic are placed in animal waterers, it can be an insect or pest repellent and help prevent worms. Do not directly feed your hens garlic because it will change the flavor of their eggs.

○ **LAVENDER** (*Lavandula* species). Most species of this popular genus are known for their beautiful purple flowers and aromatic foliage. Lavender flowers are an important nectar source for pollinating insects. This plant's essential oils are said to have a calming effect on the nervous system. Lavender is also widely used in perfumes and soaps, as a culinary herb, and as an antiseptic for cleaning wounds and as a cleansing bath. Use the leaves and flowers in nesting boxes for fragrance and as a calming agent for broody hens or birds that are stressed. Hardy to Zone 5, depending on species.

○ **NASTURTIUM** (*Tropaeolum majus*). Nasturtium is a popular garden annual known for its brightly colored flowers. The whole plant is edible, and is a good companion plant. Nasturtium does attract black aphids,

---

OPPOSITE: Comfrey grows under a fruit tree.

which worsens with drought stress and can make the plant unsightly. The seeds can be used as a de-worming agent for poultry, so collect them for that purpose.

○ **NETTLE** (*Urtica species*). Stinging nettles are not exactly a gardener's best friend, but can be a useful plant for poultry and for us as well. This perennial is best known for its stinging hairs, which cause an irritating skin rash. Nettle is one of the richest sources of chlorophyll in the plant kingdom and is a dynamic accumulator. It is a good compost activator, and as a companion plant it is said to make nearby herbs more potent and productive. Known for fattening up birds, nettle increases yields in egg production and even milk production in cows. To use this plant, harvest it (ideally with gloves and long sleeves) and dry the stems and leaves, which will remove the sting. Chickens may like the dried parts cut up, but many animals will eat nettle stems and all. Leaves can be boiled and used as a tea for us and for chickens, and when mixed with mash they can be a good wormer. Dried nettles can be powdered finely for later use. If this plant is not already on your property, you can find it in seed catalogs.

○ **RUE** (*Ruta graveolens*). This strongly scented hardy perennial is a common garden plant. Rue has a long history of medicinal and insecticidal use, and is most known for containing rutin, a substance that strengthens the eyes. The leaves of this plant contain a sap that can cause skin irritation, so it should be handled carefully. Rue can be an effective lice repellent if the leaves are dried and turned into a powder, and sprinkled in a dust bath area so the chickens get it under their feathers. A border of rue can repel insects, rabbits, and sometimes even dogs and cats. Hardy to Zone 6.

○ **WORMWOOD** (*Artemisia absinthium*). This hardy perennial is a bitter plant and has toxic properties if ingested in large quantities or for long durations. Wormwood is best used as an insecticide by making an infusion from the leaves and using it to wash for lice, mites, and other insect pests. Its roots are not ideal for companion planting because they excrete hostile exudates, making nearby plants struggle. After a few years, the plant can become leggy and benefits from shearing. 'Silver Mound' is a common ornamental. Hardy to Zone 4.

 POISONOUS PLANTS

As a landscape designer, I am often asked by concerned parents and pet owners to make sure there are no poisonous plants in their garden design. During this conversation, I can usually point to several plants in their existing landscape and talk about their toxic qualities. Many plants have a property that might make them qualify for the toxic plant list but the likelihood of poisoning humans or pets is very low. For example, apple seeds contain a cyanide compound called amygdalin, and if consumed in large quantities can prove fatal, but it is not likely that a person or an animal will ingest lethal quantities of apple seeds.

Plants have particular properties to defend themselves, because they can't just run away if they are being eaten. While some have thorns as protection, many plants have toxins in their stems, leaves, or roots that taste bad to insects and mammals.

➡ **CAUTION:** Many factors affect whether a plant would be poisonous to us or to chickens and other pets: different parts of the plant can be poisonous: roots, leaves, or berries. The amount of the toxic plant part consumed makes all the difference, and toxicity varies with different plants. The weight of the animal consuming the toxic material is also relevant.

Experienced free-ranging chickens are good at not eating anything poisonous. Most healthy animals learn fairly quickly what is good to eat and what is not. It is almost as if they have a sixth sense. But not all animals have that sense if they are underfed or not acclimated to being outside of their confined spaces. Most animals also have an instinct for eating specific plants that help treat health issues, which has been imprinted in them over time during their evolution as a species. A chicken might eat a poisonous plant because it is hungry, it has few choices for foraging, it is bored, or it is inexperienced at foraging.

tip

Do not feed your chickens yard trimmings from plants you aren't sure about and which could be poisonous.

growing worms for chickens

Our household has two different worm factory systems. It is easy to do, low maintenance, and is good for the environment. With a worm bin, you can decompose food scraps, which will turn into rich vermicompost for your plants. As the worms multiply, feed them to your chickens as a source of protein. Using a worm bin for waste keeps it from going to landfills and reduces the amount of methane released into our atmosphere.

Worm bins are a win-win system for gardeners, especially chicken owners. You can feed kitchen scraps to your birds, or you can feed the scraps to your worms instead, creating good soil quickly and providing a highly nutritious food source for your flock. By not feeding your chickens certain kitchen scraps, you also prevent them from developing bad habits in the garden, such as going after prized food plants. And you can give the worms the processed food waste that chickens wouldn't normally eat, like breads or pastas, which is like chicken junk food.

○ **RED WORMS** are used in worm bin compost systems, and can be purchased from feed stores and some retail nurseries. Bins can be purchased ready to go, complete with starter worms, or they can be created easily with any kind of sturdy box with a lid; you could even use a Rubbermaid tub with a lid with holes drilled throughout the sides and top for ventilation and in the bottom for drainage into another box or tray. To start, simply layer the container with at least 12 inches of bedding (moist, shredded paper or leaves work well), then the worms, then add your food scraps, and top it off with more bedding and a piece of cardboard. Worms like the dark. The composting, worm-breeding process starts slowly, but soon you will have a growing worm community producing some of the best fertilizer on earth. Just be sure to keep the box in temperatures above freezing. Many people keep the bin in the garage, especially during the winter.

○ **MEALWORMS** are another worm you can grow at home but in a different system, more like raising crickets. Take an old fish tank or plastic bin with a wire mesh lid, and add 1 to 3 inches of rolled oats, wheat bran, or corn meal as bedding, which will also feed them, and then add the mealworms. To give them water, place potatoes, carrots, or apples halved and face down in the bedding and they will drink away. Be sure to keep the tank clean of any rotting or molding pieces of food, dead worms, larvae, and skins. The worms pupate into larvae, and once they mature into beetles, they will mate and make more mealworms. For the highest return, separate the beetles for breeding to prevent them from eating the eggs, and keep the system in a constant 60°F to 70°F temperature. The whole life cycle will take several months to get going. You can get mealworms at pet stores and feed stores.

plants toxic to chickens

- **BANEBERRY** (*Actea rubra*)
- **BLADDERPOD** (*Glottidium vasicarium*)
- **CASTOR BEAN** (*Ricinus communis*)
- **CORN COCKLE** (*Argostemma githago*)
- **DEATH CAMAS** (*Zigadenus* species)
- **DEATH CAP, MONKEY AGARIC, PANTHER CAP, DEATH ANGEL MUSHROOMS** (*Amanita* species)
- **ERGOT** (*Claviceps* species)
- **GREEN CESTRUM** (*Cestrum parqui*)
- **JIMSONWEED, THORNAPPLE** (*Datura stramonium*)
- **INKWEED, POKEWEED, FALSE HELLEBORE** (*Phytolacca octandra*)
- **MEXICAN POPPY** (*Argemone mexicana*)
- **MILKWEED** (*Asclepias* species)
- **MONKSHOOD, WOLFBANE** (*Aconitum napellus*)
- **MOTHER OF MILLIONS** (*Bryophyllum delagoense*)
- **OLEANDER** (*Nerium oleander*)
- **PETTY SPURGE** (*Euphorbia peplus*)
- **POISON HEMLOCK** (*Conium maculatum*)
- **ROSARY PEA** (*Abrus precatorius*)
- **SHOWY RATTLEBOX** (*Crotalaria spectabilis*)
- **TOBACCO** (*Nicotiana* species)
- **WATER HEMLOCK, COWBANE** (*Cicuta* species)
- **WHITE SNAKEROOT** (*Eupatorium rugosum*)
- **WHORLED MILKWEED** (*Asclepias verticillata*)
- **YEW** (*Taxus cuspidata*)

➡ CAUTION: Many poisonous plants listed online or in books may not necessarily be the most relevant for free-range chickens.

OPPOSITE, TOP TO BOTTOM: Chickens in confinement should not be fed any poisonous plant parts. In a diverse garden, though, chickens will tend to avoid plants with poisonous properties, such as this honeysuckle, which has berries that are dangerous only if consumed in large quantities.

What is poisonous to us or to other animals like cattle may not be poisonous to chickens, and, in fact, the plant could be just a minor irritant or even be a good food source for the birds. Poultry veterinarians who perform necropsies (postmortem exams) report that chickens rarely die from eating a poisonous plant unless the material is fed to a confined bird by an unknowing owner.

I interviewed a veterinarian who said that while working in California, she had only a few cases of chickens being poisoned by plants—oleander, tobacco, and amanita mushrooms. The plants listed here have been documented as specifically toxic to chickens and as causing some related illnesses, so they should be avoided in chicken gardens. Keep in mind that for each listed plant, there is no known scenario. The material could have been in commercially grown feed for large animal production or could have come from a garden that had very few plant options. Many of these plants are weeds found around the globe.

Of these plants, the ones commonly found in gardens are monkshood, oleander, tobacco, and yew. In my garden, I have planted some popular garden plants that may be labeled as poisonous for chickens, such as daffodils, hellebore, honeysuckle, hydrangea, and tulips, and seen my chickens completely avoid them, making them ideal additions to the garden. Just be sure to not feed these plants to confined birds. Other plants that shouldn't be eaten by chickens are rhododendrons, foxglove, and crocus. There are many conflicting reports of plant-specific incidents: for example, one chicken owner watched her chicken eat rhubarb leaves, which are poisonous, and her hen died soon after; then another chicken owner's birds loved to eat rhubarb leaves so much that the plant needed protection and those birds showed no ill effect. So, are some chickens more immune to toxicity or did the hen that died have something else going on? Maybe the timing of the leaf's age was the culprit, since toxins may increase over time. As chickens live in more backyards, hopefully more relevant research will become available.

CHICKEN-RESISTANT PLANTS

What about beautiful plants for our gardens that are chicken-resistant? Chickens simply avoid many ornamental plants. Of course, there will always be exceptions with animals, because they have different taste preferences and attitudes. Chickens usually do much more damage with their feet than with their beaks, and often they ignore plants altogether. As we do with our children and dogs, we need to supervise our chickens, especially when they are new to our garden, because some plants need protection and we won't always know which plants those will be.

BELOW: Chickens don't seem to bother daylilies.

If you have a reasonable amount of chickens for your space and a well-designed and managed garden, there are many plants the chickens may ignore. Plants often have certain attributes that make chicken less likely to destroy them, including spiky or abrasive plants that are uncomfortable to walk on, such as junipers; strong odors from plants, like herbs such as rosemary and sage; highly durable plants with strong stems and leaves.

tip

When adding chicken-resistant plants to your garden space, make sure you start with a good-sized plant and use a barrier protection method while the roots are getting established for at least one season.

chicken-resistant plants

shrubs

- **ANDROMEDA** (*Pieris* species)
- **AZALEA** (*Rhododendron* species)
- **BARBERRY** (*Berberis* species)
- **CALIFORNIA LILAC** (*Ceanothus* species)
- **COTTON LAVENDER** (*Santolina chamaecyparissus*)
- **EUONYMUS** (*Euonymus* species)
- **EVERGREEN FERNS** (*Polystichum* species)
- **FATSIA** (*Fatsia japonica*)
- **FORSYTHIA** (*Forsythia* ×*intermedia*)
- **HEAVENLY BAMBOO** (*Nandina domestica*)
- **HEBES** (*Hebe* species)
- **HONEYSUCKLE** (*Lonicera* species)
- **LAVENDER** (*Lavendula* species)
- **LILAC** (*Syringa* species)
- **MAHONIA** (*Mahonia* species)
- **MEXICAN ORANGE** (*Choisya ternata*)
- **OSMANTHUS** (*Osmanthus* species)
- **PINE** (*Pinus* species)
- **PITTOSPORUM** (*Pittosporum* species)
- **ROSE** (*Rosa* species)
- **ROSEMARY** (*Rosmarinus* species)
- **SAGE** (*Salvia* species)
- **SALAL** (*Gaultheria shallon*)
- **SPIRAEA** (*Spiraea* species)
- **SPURGE** (*Euphorbia* species)
- **VIBURNUM** (*Viburnum* species)
- **WEIGELA** (*Weigela* species)

perennials

- **BEE BALM, BERGAMOT** (*Monarda* species)
- **BLACK-EYED SUSAN** (*Rudbekia* species)
- **BLUEBEARD** (*Caryopteris* species)
- **CALLA LILY** (*Zantedeschia* species)
- **CAPE FUCHSIA** (*Phygelius capensis*)
- **CATMINT, CATNIP** (*Nepeta* species)
- **CHIVES** (*Allium schoenoprasum*)
- **COLUMBINE** (*Aquilegia* species)
- **CONEFLOWER** (*Echinacea rubra*)
- **CROCOSMIA** (*Crocosmia* species)
- **DAYLILY** (*Hemerocallis* species)
- **GOLDENROD** (*Solidago canadensis*)
- **GRAPE HYACINTH** (*Muscari* species)

- **HARDY FUCHSIAS** (*Fuchsia* species)
- **HARDY GERANIUM** (*Geranium* species)
- **IRIS** (*Iris* species)
- **JAPANESE ANEMONE** (*Anemone japonica*)
- **LADY'S MANTLE** (*Alchemilla mollis*)
- **LEMON BALM** (*Melissa officinalis*)
- **LILY OF THE NILE** (*Agapanthus* species)
- **MISSION BELLS** (*Fritillaria* species)
- **PEONY** (*Paeonia* species)
- **PEPPERMINT** (*Mentha* ×*piperita*)
- **RUSSIAN SAGE** (*Perovskia* species)
- **SEDUM** (*Sedum* 'Autumn Joy')
- **SHASTA DAISY** (*Chrysanthemum maximum*)
- **SPEARMINT** (*Mentha Spicata*)
- **YARROW** (*Achillea millefolium*)

groundcovers

- **BERGENIA** (*Bergenia cordifolia*)
- **BISHOP'S HAT** (*Epimedium* species)
- **CANDYTUFT** (*Iberis sempervirens*)
- **CHRISTMAS ROSE** (*Helleborus* species)
- **COTONEASTER** (*Cotoneaster* species)
- **DEADNETTLE** (*Lamium* species)
- **FEVERFEW** (*Tanacetum parthenium*)
- **GEUM** (*Geum* species)
- **JAPANESE SPURGE** (*Pachysandra terminalis*)
- **JUNIPER** (*Juniperus* species)
- **LAMB'S EAR** (*Stachys byzantina*)
- **LEADWORT** (*Plumbago auriculata*)
- **LILYTURF** (*Liriope* species)
- **MARJORAM** (*Origanum majorana*)
- **OREGANO** (*Origanum vulgare*)
- **PHLOX** (*Phlox* species)
- **SEDGE** (*Carex* species)
- **SEDUM** (*Sedum* species)
- **ST JOHNS WORT** (*Hypericum perforatum*)
- **SWEET WOODRUFF** (*Galium odoratum*)

annuals

- **BORAGE** (*Borago officinalis*)
- **LOVE IN A MIST** (*Nigella damascena*)
- **POT MARIGOLD** (*Calendula* species)

COLORFUL SEASONAL PLANTS

While I believe that function comes first in a chicken garden, there is every reason to include aesthetically pleasing plants. Most have beneficial factors such as being a pollination source for insects or even providing erosion control. Every season brings about change, and while some plants start to fade, others shine. A beautiful, well thought-out landscape design will consider every season and the plants of particular interest during that time period.

○ *Winter.* When I begin designing an ornamental garden, I start with winter for structural and aesthetic reasons. Winter is the most difficult season to have interest in the garden, whether it be color or texture, but a garden should always have a good winter structure. In the Northwest where I live, it sometimes seems like half of the year is winter. Since herbaceous perennials will be dormant and deciduous trees will have shed their leaves, our free-range chickens will need other cover available in those days. Dense evergreen shrubs can make good shelter and can provide a place where the birds can get out of bad weather. In this season, there is more room for the chickens to scratch around in the soil, so the plants that are around better be tough. Bergenia (*Bergenia cordifolia*) is a proven evergreen groundcover in my garden, as is the perennial hellebore (*Helleborus* species), which blooms when nothing else is happening.

○ *Fall.* Fall is the next challenging season, regarding structure and aesthetic factors. Fall is when we put the garden to bed for the winter, and the chickens get to work cleaning up after the busy growing season. For color, there are many plants that provide interest in the fall and are safe for chicken gardens. Self-seeding perennials such as Japanese anemone look beautiful, *Sedum* 'Autumn Joy' is in its glory, and *Aster* ×*frikartii* 'Mönch' is a favorite. Most deciduous trees and shrubs will have colorful foliage during the autumn months. The leaves of blueberries and the leaves and stems of some maples are vibrant.

○ *Summer.* For the long, warm days of summer, there are endless options for plants. While most of us are tending to our vegetable beds, there are certain plants that need no attention but add interest, texture, and color to our gardens. Hydrangeas offer bold color throughout the summer, while daisies (*Rudbeckia*) and coneflowers (*Echinacea*) hold strong color well into fall and provide seeds for birds. Ornamental grasses start to bloom. A couple of my favorite perennials are mallow (*Lavatera* species), and *Verbena* 'Homestead Purple', which seem to have never-ending blooms.

○ *Spring.* Gardeners and chickens alike await new growth to emerge from the soil in spring. Care must be taken to protect tender plants like hostas that come up slowly. My personal favorites that look good when there are still freezing temperatures include *Euphorbia*, whose blooms glow in early spring, as well as the flowers of many naturalizing bulbs such as *Muscari* and *Narcissus*. Later in the spring, my personal favorites are just about any of the daylilies and hardy geraniums. *Geranium* 'Rozanne' is a blue-purple flowered variety that goes full throttle from May to first frost.

NATIVE PLANTS

If you are looking for plants that require minimal work, consider using plants native to your region. The foundation of our local ecosystems is the native plants that have evolved with the regional animals and organisms to create a complex web of life and a sustainable community. These plants have adapted to the specific climate and soil types of your region, making them low maintenance and drought tolerant once they are established. Native plants provide habitat for wildlife and pollinators, and can be excellent chicken habitat.

With chickens, before buying and installing any plant, even a native plant, it is good to check whether the plants are durable enough and can provide some benefits for the birds. There may be some native perennials that are sensitive or difficult to establish, so chickens scratching around them may not work, while others will make a durable, attractive addition to your garden. My gardens contain a large percentage of native plants, mostly in the trees and shrub layers, and I leave it to tough, dependable non-native perennials for color and seasonal interest.

OPPOSITE, CLOCKWISE: Bergenia is a good groundcover for a chicken garden. Aster is a dependable fall-blooming favorite. *Echinacea purpurea* 'Magnus'. *Geranium* 'Rozanne'.

PLANTS FOR FRAGRANCE

Even if you have a well-maintained chicken coop with no lingering odor, you may want to add fragrance in your garden in the off chance that you are unable to be diligent about your coop chores. With the possibility that neighbors may be concerned about odors, you can ensure them that you are planting fragrant plants nearest their property to prevent issues. There are many plants known for their fragrance.

durable, fragrant shrubs for chicken gardens

○ **DAPHNE** (*Daphne* species). Daphne is a small shrub grown for its fragrant flowers. It has poisonous berries and dislikes root disturbance, so keep hungry or scratching chickens away. Zones 6–9, depending on species.

○ **HONEYSUCKLE** (*Lonicera* species). This genus of vines and shrubs has sweetly scented flowers. Zones 4–10.

○ **LAVENDER** (*Lavandula* species). These small shrubs have aromatic leaves and flowers. Zones 5–10, depending on species.

○ **LILAC** (*Syringa* species). These deciduous shrubs produce fragrant flowers in the spring, and some varieties are more fragrant than others. Zones 2–9.

○ **ROSES** (*Rosa* species). This large genus includes deciduous shrubs and climbers with scented flowers. Zones 2–10.

○ **SWEET BOX** (*Sarcococca* species).. These evergreen shrubs and groundcovers produce an aromatic flower in late winter or early spring. Zones 5–8.

○ **VIBURNUM** (*Viburnum burkwoodii* and *V. carlesii*). Many viburnums have fragrant flowers, but these two species are outstanding. Zones 5–8.

○ **WITCH HAZEL** (*Hamamelis* species). This large shrub or small deciduous tree has wonderfully fragrant flowers in winter. Zones 5–8

PLANTS AS NOISE BARRIER

Sometimes for a noise barrier and a bit of privacy, we utilize plants with rustling leaves. With chickens, neighbors may express some concern about clucking noises coming from your garden. A water feature is a sure way to muffle noises and can benefit your chickens by providing fresh circulating water.

plants to muffle chicken chatting

○ **BAMBOO** (*Phyllostachys* species, *Fragersia* species). The leaves of bamboo plants rustle in the slightest breeze. The bigger the bamboo, the more sound it makes. Be careful not to have an uncontained runner bamboo, or that might give your neighbor something new to complain about. Hardy to Zone 5, depending on species.

○ **LOVE-IN-A-MIST** (*Nigella damascena*). This self-seeding annual has bright blue flowers, which form balloonlike seedpods that rattle in a breeze. Chickens may scratch up newly emerging seedlings, so sow them in an out-of-reach area.

○ **MAIDEN GRASS** (*Miscanthus* species). Many *Miscanthus* species are available to gardeners. The large varieties create a peaceful rustling sound in the wind. Zones 4–9.

> **tip**
>
> *Every year, cut* Miscanthus *plants down in late winter and use the dried grass as bedding.*

○ **QUAKING ASPEN** (*Populus tremuloides*). This tall deciduous tree is known for leaves that tremble in the wind. It is fast growing and upright, which makes it a good choice for spaces that need vertical screening. It can also run, but is easily controlled by removing any new sprouts. Swedish aspen (*P. tremula* 'Erecta') is even more columnar. Zones 1–8.

○ **QUAKING GRASS** (*Briza media*). Also known as rattlesnake grass, this small deciduous grass has rattlesnakelike inflorescences that rustle in the wind. Hardy to Zone 4.

OPPOSITE: Variegated *Sedum spurium* 'Tricolor' comingles with chartreuse *Sedum rupestre* 'Angelina', making a dense groundcover mat.

innovative

CHICKEN HOUSING

Hen houses, chicken coops, or night shelters are a must for every flock. If they are well designed, these little structures can be fun, colorful, and add an attractive design element to any garden. With some valuable information, you can design and build a customized coop of your own.

urban homestead

THERESA LOE, OF LOS ANGELES, CALIFORNIA, has a ¹/₁₀-acre lot and grows food on every possible square inch. Theresa is a mother of two, a garden communicator, a freelance writer, a popular blogger, and an associate producer of a nationally televised gardening program. Her training as a horticulturalist and certified master preserver keeps this urban homestead growing and preserving food to feed this family of four year-round.

Their three chickens are family pets, and free-range for a few hours every day during the week and all day on weekends. The chicken coop is a renovated doghouse with a simple green roof and an attached chicken run. The green roof, which is insulated with plants in nursery trays installed over a waterproof liner, helps regulate the coop's interior temperature, filters the rainwater, and looks beautiful.

Theresa also grows food for her chickens, including wheatgrass in several window boxes, which she rotates as the birds graze them down. The chicken manure is used as fertilizer, and barriers such as temporary fencing or a chicken tractor, made from repurposed materials, are used to protect vulnerable food plants at certain times of year. Theresa proves that having an urban edible garden and chickens can work well together.

PAGE 146: Theresa Loe's coop has a green roof. OPPOSITE: Theresa Loe and her hen Penny, a Plymouth Barred Rock.

A COMMON CONCERN OF NEIGHBORS WHO ARE ANTICHICKEN is that they think the presence of a flock will lower their property values. Indeed, a coop that hasn't been designed properly can be unsightly and cause problems other than just being an eyesore. If you haven't designed the coop with maintenance and cleaning in mind, you may end up neglecting it and it could smell bad, compared to a coop set up for easy cleaning. If a coop is not rodent proof, the structure can attract pests like rats that can take up residence with the chickens.

These are all valid concerns, but they can be avoided if the coop is thoughtfully planned and built well. So instead of settling for a chicken shanty, why not build a cute, stylish coop that is worthy of your chickens and your garden? There are many things to think about before sitting down at the drawing table or shopping for supplies. If you follow simple design steps, you will be sure to have a coop that is functional for you and for your chicken's needs, and you will be proud of it.

 ## SPACE REQUIREMENTS

I've often heard people say, "I don't have enough room for chickens." Usually, they are in an average urban lot of about 6000 square feet, and there is a legal limit of having three birds. My response is, "Why not?" You don't need a huge amount of space unless you plan on having lots of birds. I've met chicken keepers with successful free-range chicken systems in beautiful backyards as small as 800 square feet. But nevertheless, your initial planning should be focused on the footprint of the coop design.

I have read many simple equations regarding number of chickens to available coop space, ranging from 1 hen needs from 3 square feet, to 1 hen needs 10 square feet, and I wonder where these numbers come from. Figuring coop space per chicken really depends on how big your chickens are and how often you allow them outside in a run or to free range. The more free ranging or confined ranging you allow, the smaller the coop

needs to be. At night, chickens take up very little space in the coop. They perch on a bar and fall asleep. If your chickens are only using the house at night, then they may need a smaller square footage.

I recommend at least 10 square feet per bird. If a bird is living in a coop with a run attached, it should have an absolute minimum of 10 square feet total, 5 square feet inside and 5 square feet outside. An attached chicken run should be as big as your space allows. The smaller the area, the more the chickens will suffer from pests and social struggles.

 ## MOBILE, MODULAR, OR STATIONARY COOP?

A chicken coop can be a permanent fixture in your garden or it can be a mobile or modular device. Occasionally you may want to move the coop into different areas of the backyard. Or perhaps you may want to think ahead if you ever have to move: Will your coop design fit through the garden gate and be transportable? You can also make your coop modular to help accommodate a possible move, or you could create it with wheels that can be removed. With a stationary coop, you will have more choices of materials and the weight will not be as much of an issue.

> **tip**
>
> *Think about what fasteners you use to build the coop. For example, screws may cost more and take longer to build with, but they allow for easy disassembly if you need to take apart the coop.*

OPPOSITE, TOP TO BOTTOM: Jennifer Carlson's modular coop comes apart in three pieces. The roof was designed to have a steep pitch so predators would be less likely to hang out, stressing the flock. A detachable wheel on a chicken tractor makes it easy to move.

Expansion Possibilities

Once you have chickens, it is hard to stop accumulating them—almost like an addiction. It is best to plan for a larger space than you need, just in case, or to have the ability to add on more space.

Height

The height of the coop is an important design factor when planning your hen house. The coop should be tall, mostly for your convenience: you don't want to clean your chicken coop all hunched over. A taller coop has perks for the chickens too: it allows the chickens to be up off the ground, high on their roosts, which makes them feel safer, and allows more air circulation inside the coop.

Door Locations

How many doors? How big? A human door is a must, so you can easily access the coop and perform regular cleaning and maintenance. If the coop is small, that door can be the roof itself, on hinges, or an entire wall. Your chickens need to have their own doors, which should be about 10 inches tall by 10 inches wide. Chicken doors should be on hinges, so during the day they can be left open and during the night they can be shut for safety and security. An open door on a hinge can double as a ramp for the chicken to easily enter and exit the coop.

Floor Plan

If you are designing the coop from scratch, it's best to start by drawing out the floor plan of the coop. Locate the walls, ventilation, roost bars, nest boxes, water, food, ladders, windows, people doors, chicken doors, and any other essential features.

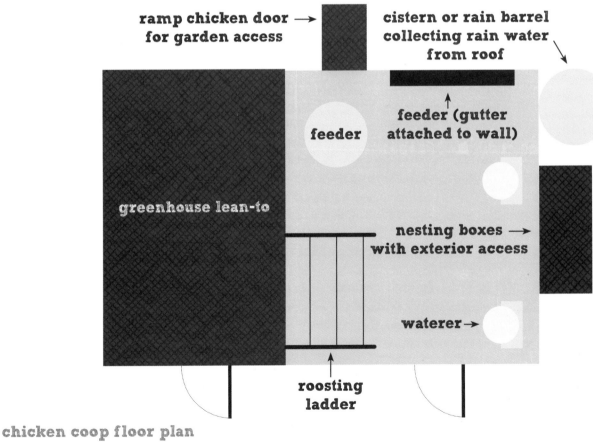

ramp chicken door → for garden access

cistern or rain barrel collecting rain water from roof

feeder (gutter attached to wall)

feeder

greenhouse lean-to

nesting boxes → with exterior access

waterer →

roosting ladder

**chicken coop floor plan
also see elevation drawing, page 164**

For the chickens' safety, coops need to provide a few basic necessities: protection from the elements or extreme weather, protection from predators, and good ventilation. These elements should be first and foremost when designing and building your coop.

Protection from the Weather

Chickens need to have shelter from the wind, rain, snow, and heat. A sturdy, tight roof on the coop will help ensure that the rain won't get in from above, but also consider and plan for the occasional sideways rain that can come in through a large vertical opening. Chickens can be affected by wind chill just like we can, so figure out the direction of the prevailing winds in that spot and plan to create a wall or plant shrubbery as shelter near openings. If snow is a problem in your area, you may want to slope the coop roof steeply and make sure that coop openings can't get buried.

Protection from Predators

Persistent predators will try to pry their way in through doors and windows of the coop, so you will want to make sure the locks and hinges are secure and too complicated for little paws to open. Predators will dig and crawl under the coop, making your flooring choices important. They will even chew through wood to get at eggs or chickens. Knowing the predators in your area will help you in designing the coop, but for safety, it is best to be overprepared. Also read about fencing for more information on fencing materials and options for protecting chickens.

Ventilation

Air circulation and ventilation are important year round. Because chickens don't have little toilets to flush, their wet manure accumulates on the ground (sometimes on the walls) and saturates the air with moisture and ammonia. Just like in our homes, fresh air needs to be circulated into the space and stagnant air needs to be ventilated out. A well-designed coop will have vents in low spots of the hen house as well as in high points. Whether you install windows or openings with wire mesh or ventilation hardware, it is handy to be able to open and close hatches over them to allow for different amounts of air circulation during different times of the day, night, or year. For especially hot and cold spells, or at night, it is ideal to have the flexibility to provide your chickens with fresh air flowing through the coop. If you can smell ammonia and you see cobwebs, then your coop needs more ventilation.

when is it too hot or too cold for chickens?

Most chicken breeds will thrive in temperatures of 40°F to 85°F and can live in much colder conditions. Dehydration and heat stroke are common causes for death in hens.

○ **COLD:** Certain breeds will be hardier than others, and large combs, wattles, and feet can be susceptible to frostbite. You will see a drop in egg production when temperatures drop. But most chickens can withstand cold as long as they are in a dry, have draft-free shelter, and have other hens nearby giving off body heat.

The most important thing is to provide plenty of fresh water.

Using a tank de-icer or small pet water heater is a good way to prevent ice from forming in waterers and could save your chickens' lives. These devices can be found in pet stores and feed stores, especially during the winter. Some chicken keepers choose to use a heat lamp to keep the birds warm and as a light source to help continue egg production, but be careful: chicken coops commonly catch fire from heat lamps. You should feed the chickens in the late afternoon so their crops are full when they go to bed.

○ **HEAT:** Chickens don't sweat and will often pant when they are overheated, like a dog with their beaks open. When temperatures are constantly over 85°F, usually shade, ventilation, and plenty of fresh water are enough to keep the birds comfortable. If you have a smaller hen house or a large number of birds, you may need to add a fan in the coop to make sure air is circulating on the hottest days. If chickens are contained in a coop or tractor with a metal roof or siding, make sure they have extra fresh air since a metal structure can turn into an oven. Some breeds are better adapted to warmer climates than others.

INTERIOR DESIGN

The interior of your coop needs to contain a few must-have elements for basic good health for the hens:

o *Nest box.* One nesting box can serve 2 to 3 hens and they only need to be 1 square foot (12 inches tall, 12 inches deep, 12 inches wide) of space. My coop's nesting boxes are in a large space in one area so the chickens can roost together. Chickens prefer to lay eggs in a dark, sheltered place, preferably off the ground a little bit, but if they are too high, they might need a ladder to get to it. You can hang burlap over the front of the nesting box to give them more privacy while they are going about their business. The best coop designs have access to the nest boxes from the exterior, so you don't have to enter the coop to get the eggs. Nesting boxes can be made of many materials: a wooden box, a small pet carrier, or even a simple cardboard box (which must be replaced regularly). For years, my coop's nesting boxes were obsolete city recycling bins.

tip

Make sure your nest boxes are covered and the chickens cannot roost on the edges. Otherwise, you will be constantly cleaning manure off the eggs. A steeply sloped roof over the nesting box is a good way to prevent this.

o *Roosting bars.* Chickens like to roost on something at night while they sleep as a way to be off the ground where predators can't reach them. Each chicken should have at least 1 linear foot along the roost. So if you have six chickens, you will need 6 feet of roosting bar. In general, the bigger the chicken, the bigger the diameter of the roosting bar. Smaller breeds should have a smaller diameter roosting bar. If you live in a cold climate, you may want the chickens sleep flat footed so their bodies cover their toes to prevent frost bite, and for this case, using a 2 × 4 roosting bar (with its wide side up) works well. Perfectly uniform perches (or from materials that are plastic, wire, or sharp) are said to be a cause of bumble foot, which is an ailment that is also caused by vitamin deficiency. A roosting ladder made out of branches is a good way to provide multiple roost bars with some higher and some lower. Having roost bars at different heights is good for the different sized birds and rankings in the pecking order. If you make a roost ladder, make sure there are at least 18 inches between the rungs.

o *Bed pans.* The placement of the roost bars should not be overhead from waterers or feeding areas, for obvious reasons. Instead, have the roost bars placed over an area with removable pans, which you can slide out and clean the manure out of easily. Old plastic window box containers work well, or even old cookie sheets from thrift stores.

o *Water.* Of all the chicken chores, providing clean water ranks among the most time consuming and most important. Be sure to have the water in an area the chickens can't climb on top of, preferably off the ground. I have several waterers for my birds, including an automatic gravity-feed waterer on a cinder block and a hanging 5-gallon bucket with nipples. There are many kinds of waterers and feeders on the market—plastic or galvanized hangers, troughs, tubs—that can be found at feed stores in a variety of sizes, materials, and prices.

building a nipple waterer

Building your own chicken waterer is cheap and easy, and this design keeps the water clean. You'll need:

o **5-GALLON BUCKET WITH LID**
o **11/32-INCH DRILL BIT**
o **3 POULTRY NIPPLES (WITH RUBBER GROMMETS)**

Drill three holes in the bottom of the bucket, spaced equally in a triangle for maximum efficiency. Place a rubber grommet in one drilled hole, followed by one poultry nipple, and repeat until all three grommets and nipples are installed. Hang or fasten the bucket with a rope, chain, or cable so the bottom hangs just above the hen's height. Fill with water, and cover bucket with the lid to keep the water clean.

OPPOSITE: Rachael Vitous and Dan Bauer designed this cozy coop to match their house and built it with salvaged materials.

o *Feeders.* Chickens need to have feeders that are easy to reach and free of manure, dirt, and other debris. Similar to water, it is best to have the feeder elevated so rodents can't reach it and chickens don't roost on top of it. Hanging feeders are a good method, or having an automated system with a gravity feed. I have simple plastic troughs that are easily moved after the chickens have finished eating, and I also use a gutter that hangs low on the wall. There are many types of feeders you can make yourself, like a large-diameter PVC pipe or a 5-gallon bucket with holes drilled at the bottom and sitting in a larger flat container like a container plant pot base.

o *Food storage.* It wise to keep feed in a metal container, such as a galvanized garbage can with a tight-fitting lid. Plastic containers can be chewed through easily by rats.

🌳 OPTIONAL ITEMS

While some items aren't required for all coops, they can certainly make tending your flock easier.

o *Electricity.* Electricity in the coop makes life easier, especially in the winter months when you may need to heat your waterers, or you want to have artificial lighting inside your coop so the chickens continue to lay eggs. If you choose to put an electrical outlet inside your coop, be sure to have a GFI outlet and locate it as high as possible, to prevent it from getting wet or pooped on.

o *Artificial lighting.* In the deep winter, it is nice to have lights in the coop, if not for increased egg production, then to see what you are doing in the early morning or late afternoon hours. A simple digital or analog lamp timer is especially helpful in turning your light(s) on and off on a schedule. Chickens don't need lights on all the time, and if you choose to provide supplemental lighting for egg production, your chickens may only need a few extra hours of light per day, for a total of 14 to 16 hours. Most hardware stores carry simple, inexpensive timers. If your power goes out, the timer will probably need to be re-programmed.

o *Storage.* It is helpful to have storage either in the coop or close by for feed, bedding, and cleaning supplies. Having a spot to hang your cleaning tools inside or outside the coop will make chores much easier.

Rainwater harvesting. Having a gutter on your coop leading to a rain barrel or cistern is good for water conservation, storm water management, and a great source of nonchlorinated drinking water. Metal is the preferred roofing material for collection of water that is going to be consumed by animals or humans. If your roof is asphalt, it may have contaminants, but the water is still good for watering ornamental plants.

o *Heated waterers.* Heated waterers are definitely worth the increased expense and can save you from having to carry water and to ax your way through ice every morning during the winter. Most feed stores and pet stores have water bowl heaters for dogs and horses. They are simple devices you put inside the water to keep it from freezing or they have heated bases. You can use a heated seed mat or even a crockpot if you are in a bind, but be careful that the water doesn't get too warm.

o *Automatic door openers.* Automatic door openers can be especially handy if you travel a lot and don't always have a chicken-sitter to help let your chickens in and out while you are away. An electric system will take a certain amount of planning and installation, to make a door open and close on a timer. Manual pulleys work well to open doors from the outside.

o *Interior paint.* While it is not necessary to paint the interior of the coop, a painted surface can make cleaning much easier. Also, red mites like to hang out in the cracks and crevices of your coop, making your hens uncomfortable at night when they roost. By having a painted interior, presence of mites will be easier to monitor. There's no need to have designer color and wall treatments; just a coat or two will do the trick. Be sure to let the coop air out for a few days before locking the chickens in at night.

o *Fake eggs.* Fake eggs are nice to keep in a nesting box. They encourage chickens to lay where you want them to, because a chicken will think that if another hen thought it was a good spot to lay an egg, she will follow suit.

o *Decor.* Chicken coops can have artistic flair, and while the chickens may not appreciate it, we will. Hang a wreath on the coop door, decorate the exterior walls, add a flower box to the side, hang a trellis on a side wall, or grow a vine to soften the structure.

OPPOSITE, CLOCKWISE: This coop has pullout manure trays under the roost bars for easy cleaning. Prayer flags hang inside this coop. This chicken door in a run is opened and closed by a simple pulley system from the other side.

CHICKEN COOP STYLE

This is where you have the chance to have fun and get creative. Why not make your chicken coop a design feature or accent in your garden? You can have a stylish coop that matches the design or colors of your house and compliments your landscape. Styles can vary from the simplest utilitarian structure with a few details to a stone-clad Tudor-style chicken house complete with a chimney. Popular coop styles include everything from classic monitor barns (barns with a raised ceiling in the center that allows for overhead storage and has a tall center roofline) to a modern architect-designed metal enclosure.

I like to make the chicken coop fit with the surrounding structures. In a cottage garden, I would design the coop to match. If your house is a Craftsman, follow the same details in roofline, trim, and colors. If there is a brick façade on your house, why not repeat that on your chicken coop? Design it so it makes you happy, since you will be visiting it every day. Browse ideas from websites and take chicken coop tours in your area. Many major cities have annual coop tours where chicken owners open up their hen houses to the public for a weekend and you can see what others have done and ask questions.

MATERIALS

In choosing building materials for your chicken coop, there are many options to consider.

Roofing

Roofing can be made out of lots of materials. Metal is a long-lasting material that comes in many different colors and is ideal as a surface for collecting rainwater. Asphalt shingles are less expensive and still very durable. Plastic corrugated roofing works well too and comes in many different colors, even translucent to let in more natural light, but it has a low insulation factor. Maybe you will consider a green roof, where you can grow plants that work well in your climate. A "green roof" is a good way to insulate the roof, keeping your chickens warmer in the winter and cooler in the summer. No matter what material you choose, consider the slope of the roof and which way it drains, and consider using gutters leading to a rain barrel.

> **tip**
>
> *The roof on your coop should slope away from the run or anywhere that the chickens are, so you can avoid having a muddy mess.*

Walls

Many coop walls are made out of simple framing and some sort of plywood. If you live in a cold climate, you may want to insulate the walls to prevent heat from escaping, but be careful to design the walls so that they are rodent proof, since rats and mice love to live in coop walls. Exterior walls can also have siding to protect the coop from the elements. Painted plywood will work, but a layer of siding will add many more years to your hen house.

OPPOSITE: The clear roofing material over this coop's attached run lets in natural light and doubles as a drying rack for onions and garlic.

Floor

Of the many options for flooring for your hen house, all have their pros and cons. Maintenance and cost are the two main factors for choosing a floor material, but also consider how predator proof it will be.

o *Dirt flooring.* Dirt flooring in your coop is the least expensive, assuming you don't need to import new soil. You should only consider a dirt floor if you have well-drained soil, or it could become a mucky mess. Dirt flooring offers the least protection from predators and could harbor parasites unless you have well-managed deep litter mulch or clean often. Soil will stay cooler, which is great in the summer months, but could be too chilly for the chickens in the winter months.

o *Wood.* Wood is mentioned in most coop building references as a good flooring material, but I wouldn't recommend it unless your coop is elevated at least 2 feet off the ground and wood is the only option available for you. Rodents can easily chew through wood, and it absorbs spilled water and manure, making it a potential sanitation and maintenance issue. If a wood floor is sealed or made out of treated material, it could enhance maintenance and longevity. But keep in mind the toxicity with all treated wood, especially where chickens will be spending a lot of time. Many people who use wood place a tough laminate surface over it.

o *Concrete.* Concrete is a long-lasting, predator-proof material that is easy to clean. Having a poured concrete floor can be costly for large coops, but it can be worth the investment for your peace of mind. If you choose poured concrete, plan on having it poured with a slope toward a drain, for when you hose out the coop.

o *Concrete pavers.* My flooring material of choice for inside the chicken coop is concrete pavers. Pavers cost a bit more that poured concrete, but you have more flexibility if you need to move the coop or get underneath it for some reason. The bigger the pavers, the better: mine are 2 feet square. Be sure to properly prepare a gravel base under the pavers for stability and good drainage.

o *Gravel.* Gravel is another inexpensive material that can be a good base flooring in your coop. It is less expensive than concrete products, and provides good drainage. But it is similar to dirt floors with regard to cleaning and being predator proof. If you choose gravel, I recommend that you do not use large round gravels like cobble or washed gravel, but instead use crushed rock with "minus" in it, which is the dust from when the rock is crushed. Be sure to use a plate compactor on the gravel to eliminate any air pockets that can make it easier for predators or pests to dig underneath.

Bedding

For floor bedding material, the most commonly used are straw, wood shavings, and leaves. I rotate different materials on a regular basis. Make sure the material is dry and isn't allowed to get moldy, which can kill chickens.

For nest box bedding, it is easiest to use the same material you are using for floor bedding, but I particularly like nest box liners. They look like Astroturf and provide a soft bed for the egg, air circulation, and are easy to clean so you can reuse them. They are available in poultry catalogs and some feed stores for only a few dollars each. As gardeners, we spend a lot of time cutting back plants, and we can use many plants as bedding and nest box material. I often use dried grasses or leaves as bedding when it is available. Just be sure it is dry and free of mold.

Finding Materials for Your Coop

You may have materials lying around that could be great for building a chicken coop. Consider what you have on hand first, to reuse and recycle, and then look elsewhere for the other needed materials. Most little coop structures have odd dimensions and don't require full-length lumber, so you can often use salvaged materials.

o *Salvage stores.* Salvage stores are my favorite places to look for building supplies. They specialize in selling reclaimed building materials. The materials are more affordable than buying new materials and sometimes even have historic interest. And you can often find new materials at a salvage store, from a builder who had surplus roofing or windows. When shopping for used materials, be aware that they could potentially contain toxic materials, especially older painted wood. Lead can commonly be in old paint, which is dangerous for your chickens and to you by consuming their eggs. A good savage store will have warning labels on items with high risk.

OPPOSITE: At the end of the growing season in this vegetable garden, hens are put to work in an A-frame tractor to help clean up.

o *Thrift stores and other sources for used materials.* You'd be surprised what you can find in a thrift store. I am a bit of a thrift store junky and have found amazing items and deals that can be used for chickens, and for gardening as well. For example, an old bookshelf or Tupperware storage bins can easily be turned into nesting boxes, and ladders or broomsticks can be turned into roosting bars. Many of your coop's interior features can be repurposed household items. Another place to find used materials is the local classified ads, whether it be in your newspaper or online at sites like craiglist.org or freecycle.org.

o *Hardware stores and big box stores.* If you like one-stop shopping, this may be where you end up getting many of your materials. You can most likely find everything you need at big box hardware stores, but they may have limited options and not stock everything you are looking for. It is usually easy to special order specific items, and many of these stores offer low-cost (or sometimes free) delivery.

o *Feed stores.* Your local feed store can be a good resource for feed, advice, and some building materials. They usually carry a selection of fencing materials as well as waterers and food dispensers.

o *Online catalogs.* For specialty items like waterer nipples or incubators, you can find many options in farming or poultry mail order or Internet catalogs. See the Resources section at the back of this book for ideas.

Buying a Premade Chicken Coop

Not everyone has the tools, time, or know-how to build a chicken coop. If you are not a do-it-yourselfer, there are many ways to get a chicken coop. Start by browsing the Internet for catalogs that carry coops in the size you are looking for. Collect pictures of what you want and be sure to write a list of your must-haves versus your wish list, since some features can be added on later.

....

OPPOSITE: Nicole Starnes Taylor designed her coop-tractor to feel modern, with distinctive metal screening and painted wood.

o *Custom coop builders.* There is a growing breed of qualified builders who can build a chicken coop that will work for you. They are in every city and charge a reasonable rate for their time and materials. Builders should be able to tell you your climate requirements, and possibly the predator profiles in your area so you will have a secure coop. If you choose to work with a custom coop builder, have your budget figured out in advance. Coops can cost a few hundred dollars up to several thousand, so having this crucial factor pinned down ahead of time will save you and the coop builder time in planning.

o *Kits.* Kits are a less expensive way to buy and build a chicken coop, but will still require that you have the tools and time to put it together. Many kits are limited in the size and accessories, and could be considered for expansion if you choose to add more birds later on. You can paint kits to match your home and you can attach a bigger run. Just be sure what you are buying: if you can, ask to see an already built kit. The craftsmanship and quality of materials are important. Pressboard, for example, may fall apart after one year.

o *Do-it-yourself plans.* Many companies and individuals sell custom chicken coop plans. This may be a good approach for you if you are able to build a coop but can't picture what you want. Before paying for blueprints, be sure to research the competition and get the best plans that suit your situation. Make sure your plans include diagrams, a detailed material list with specifications, and step-by-step instructions. As with the kits, do-it-yourself plans allow you to customize the design.

Renovating an Existing Structure

If you are lucky enough to have a shed or an old barn on your property, it can be converted into a wonderful space for chickens. It just needs your vision, planning, and your execution. A garage is a possible place to keep chickens at night. But chickens create dust, which may not be ideal if your restored hot rod shares the same space.

GREENHOUSE CHICKEN COOP

A classic permaculture idea is to have your chicken housing built as a part of a greenhouse design, which can be beneficial for the chickens and the plants growing in the greenhouse. The chickens' body temperature helps keep the greenhouse warm in the winter months and regulated at night. The CO_2 released from the chicken's exhalation is absorbed by the plants and can increase the vigor and yield.

A greenhouse system can be as simple as a rudimentary hoop house (an inexpensive, arched structure with plastic covering to extend the growing season) or complex with high-tech systems for serious plant production. Your greenhouse design needs to consider the size of your intended greenhouse, how many chickens you are considering, what kinds of plants you are growing, your climate, what kind of ventilation is in place, and how your irrigation system works. The elements in your greenhouse need to be considered if you plan to add chickens. Ideally the coop would be attached to the north side of the greenhouse to allow maximum sunlight for the plants. The air circulation and exchange should be planned to provide good ventilation, especially during the warm summer months, so the chickens won't get too hot, and then to prevent the dust from the chickens from getting on the plants. And just like other chicken structures, it will need to be predator proof.

You probably won't want to use the greenhouse like a run for the chickens, because they can damage young plants. But that depends on what you are growing and what stage they are in their maturity; seedlings in trays, for instance, need much more protection than shrubs in five-gallon containers. The chicken manure is another factor in prohibiting them from free ranging inside the greenhouse. It could be difficult to clean up, and the ammonia released from it could be problematic for the plants.

BELOW: A small free-range flock forages in the eco-lawn next to this front yard garden.

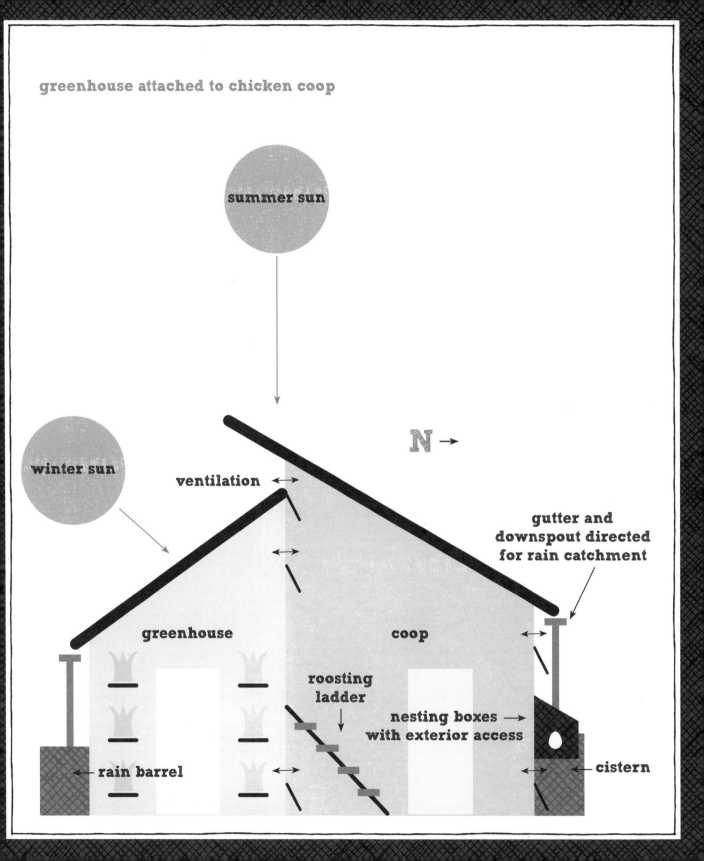

TRACTORS AND ARKS

A chicken tractor (sometimes called an ark) can be a wonderful tool for gardeners who keep chickens. A chicken tractor can be designed to fit over your garden beds or on top of raised beds, so the chickens can work in specific spots. A tractor can be used for tilling and can be rotated through a garden during different times of the year. It is even a good sod removal tool. If you want to use a tractor as a permanent home for your chickens, the tractor will need to be moved regularly, so make sure the structure is lightweight.

Chicken Tractor Style

Many chicken tractors are A-frame style or similar to mini hoop houses. But a tractor can be creatively designed to match the style of your home and to be an accent in the garden.

Factors to consider for chicken tractor design include: Are the materials lightweight enough so the tractor will be easy to move around? Does your design include easy access to food bins and waterers for refilling them? Will the chickens live in the tractor full time or temporarily to do some gardening work? How many chickens will be kept in it and how often will you be rotating it to new areas?

OPPOSITE: Ed Rains made this chicken tractor from an old 55-gallon barrel attached to a light-weight rebar frame.

○ *Lightweight framing.* Some good materials to use for the chicken tractor's framework that are both lightweight and can withstand outdoor elements are Schedule 40 PVC and rebar. Wood is used a lot as well, but the weight can add up, depending on how large the tractor is and whether it will get wet and waterlogged. Bamboo is a lightweight material, but it will eventually break down.

○ *Moving a tractor.* Know how you are going to move the tractor and when. Many farmers may have a horse or backhoe to help do the work of moving this structure, but you will probably be using yourself and perhaps family and friends to help. Consider placing wheels on one end of the structure, and make the bottom boards durable enough to withstand being dragged from place to place. Much like a snow sled, it can have a rope attached to both sides on one end so you can pull it.

tip

If you only use chicken tractors every once in a while, when chickens are not using them they can serve as large cloches or cold frames. Simply put the tractor over a bed with sensitive plants and cover it with horticultural fleece or clear sheet plastic.

 ## BROODY BOX OR SIN BIN

Sometimes we need to put a chicken in a separate structure temporarily. Having a separate little structure can come in handy for many uses. You can use it to separate a chicken from the flock that needs medical attention. You can place new chickens in it, to gradually acclimate them to the flock. You can give a broody hen a place where she can sit on her eggs undisturbed. It can act as an attached modular unit for a chicken tractor. It can house a troublemaker in the event that the chicken needs to be temporarily separated from the flock.

Any chicken placed in a small space should have adequate water, food, and air circulation, and should be able to get outdoor exercise at least once a day. This type of structure can be made in many shapes, but I suggest making it between 5 and 10 square feet.

for fun: chicken bingo

A chicken bingo box can be a fun addition to your chicken accessories. Bring it out at backyard barbeques for entertainment and some innocent gambling or even use it for fundraising events to bring in money for a good cause. So, what is chicken bingo? The game and rules are easy: On the bingo mat, you have square grids with random numbers placed. Use 1 through 50 or an appropriate quantity of numbers for the space. All participants take a number or two (for a price, if this is used for fundraising or backyard betting). Put the chicken in the large box on the mat with some food and water, and wait for her to poop on a number. The number that gets pegged first wins.

The structure of the bingo box can be as simple as netting around a driveway covered in sidewalk chalk numbers, or a labeled mat that slides under a wire transportation crate. If you plan on it on a regular basis, consider making a modular box that can be used for storage or possibly as a cold frame similar to a tractor when it is not in use.

GEODESIC CHICKEN DOME

A dome-shaped, lightweight coop and run can be moved around much like a chicken tractor. You can make a geodesic dome out of many materials—PVC, metal, or wood—and then wrap it in fencing. There are even inexpensive playground climbing structures that would work well as a frame, or old dome tent poles can be used and reinforced. A modular predator-proof night house, close to the edge of the dome, will help ensure easy access for feeding, watering, and collecting eggs. With a chicken dome, instead of using rectangular raised beds, use keyhole-shaped beds, which you can plant in rotation.

geodesic chicken tractor

OPPOSITE: Geodesic chicken tractor and keyhole garden beds.

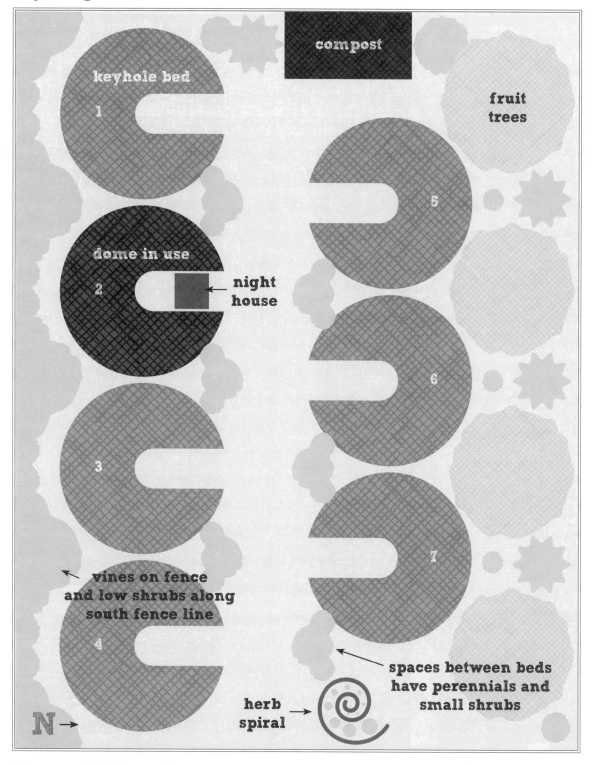

keyhole garden beds

keyhole bed

1

compost

fruit trees

dome in use

2

night house

5

3

6

vines on fence and low shrubs along south fence line

4

7

herb spiral

spaces between beds have perennials and small shrubs

N →

a lightweight chicken tractor

This lightweight chicken tractor was designed to fit over 4-by-8-foot raised beds. Its wood top bar overhangs on both ends to serve as a handle for moving the tractor. To make this tractor, you will need:

- 3, 2 × 4 × 8-FOOT CEDAR BOARDS (FRAME)
- 2, 2 × 2 × 8-FOOT CEDAR BOARDS (DOORS)
- 5, 8-FOOT × 1/2-INCH SCHEDULE 40 PVC (HOOPS)
- 1, 10-FOOT × 5/4-INCH CEDAR DECKING BOARD (TOP BAR AND HANDLE)
- 30 FEET OF 3-FOOT FENCING (3 × 3 INCH WELDED METAL MESH)
- 1 BOX SMALL HORSESHOE NAILS
- 1 BOX 2-INCH ROOFING SCREWS WITH WASHERS
- 2, 3-INCH EXTERIOR HINGES
- 2 EXTERIOR GATE LATCHES
- 1 SMALL TARP

The framework arcs can be made of rigid PVC or rebar with fencing attached; optional support triangles can be attached at the ridge, depending on the size of the tractor. A tarp is bolted over a portion of the tractor and rolled up when not in use, and a door on each end can be for access or a portal to a broody box used for shelter. When the tractor is not in use, it can be covered in Remay cloth and placed over veggies to extend the growing season.

OPPOSITE: Bandit walking back to the coop in the Fries garden. PAGE 172: Lee Reid built this adorable detailed coop, which even has a stone chimney. PAGE 173: Alan Farkas designed this modern coop to match the house, with aluminum grating for the sides, roof, and flooring, stained cedar for the nesting boxes, and cement board enclosures where the chickens roost. PAGE 174: This coop was placed in an area of the garden that is visible but softened by plants. PAGE 175: This adorable custom coop designed by Keith Forde is a cottage style. PAGE 176: Heather Bullard's charming coop, Chez Poulet, with French details, is painted inside and out and has built-in storage under the nesting boxes that are accessible from the outside. PAGE 177: Berg Danielson's small Island Coop works well in a small urban garden.

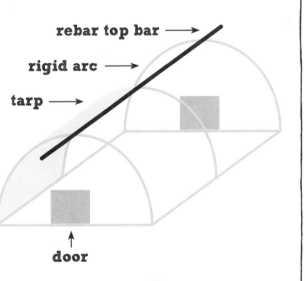

rebar top bar →

rigid arc →

tarp →

↑
door

friends and foes

OF HENS IN THE GARDEN

Chances are, if you are interested in keeping chickens, you already have experience with other pets. I think of chickens as the gateway animal in urban homesteading: once you start with chickens, it is hard to stop accumulating more of these birds, as well as bees and maybe even other farm animals. Bringing chickens into your backyard adds a new element to the ecosystem of your garden. So we may need to prepare our existing pets for this change, as well as plan for any unwelcome animal visitors.

the collectors at

old goat farm

FORMER CITY DWELLERS GREG GRAVES AND GARY WALLER found the perfect idyllic farm to call home at the base of majestic Mount Rainier, where they are surrounded by lush gardens and a plethora of animals. Dogs, goats, ducks, turkeys, peacocks, guineas, and over sixty chickens all call Old Goat Farm home.

Their hen house is an older building where all of the birds share space under one roof. Next to it is a large vegetable garden that the birds have access to in the off-season. For the remainder of the year, they free range in an area with native plants that includes a pond, which is slowly being renovated to have trails throughout. Greg and Gary love plants as much as animals, and have a large collection of specialty and rare plants from all over the world through travel and friends.

The soils at their farm have been a challenge, but you wouldn't know it. The expansive gardens are beautifully designed and planted, and even include artwork and memorials. The chickens don't have access to those gardens, but their manure is used for fertilizer after it has been passively composted.

Old Goat Farm also has a specialty retail nursery that hosts gardening groups and teaches workshops. Most of the chickens there have been given fantastic names, like a rooster named Gregory Peck and a group of hens all named after First Ladies.

PREVIOUS PAGE: Inside the gate, hen Loretta Young gets special treatment after being hit by a car: she is allowed to forage in the vegetable garden at the peak of the growing season. OPPOSITE: Greg Graves and Gary Waller in their vegetable garden.

PREDATORS AND PESTS

Chickens are a prey animal, and so they have eyes on the sides of their head, enabling them to watch for predators all around. They are alert to predators but have very few defenses, making them easy targets. Many chickens that free range are keenly aware of overhead predators and will duck under cover in shrubs or brush if a large bird is overhead—or an airplane, for that matter. In the wild, there is a balance of predator versus prey, and chickens are near the bottom of the food chain. Most predators of chickens also control the populations of other animals that we may consider pests, like mice, rats, voles, rabbits, and squirrels. Predators are just doing what they naturally do to survive and help keep the ecosystem in balance.

To best protect your hens from predators, get to know and understand the animals in your area. Start by asking neighbors and other chicken owners. There are two main kinds of predators to worry about with chickens: land-based and air-based predators. Chickens in confined spaces are most susceptible to land-based predators such as raccoons and foxes, which can climb over fences, into the coop, and dig and crawl under fences to get to the birds. Free-ranging chickens foraging in the open may be most susceptible to air-based predators, such as raptors like hawks and eagles. Many indicators identify the presence of predators: feces or scat, tracks, claw marks, or chewing on wood and trees. If you see signs of a predator trying to pry its way into the coop or breaking apart your fencing, act immediately. Some animals are persistent and will keep coming back until they get what they are after. Any delays could mean the end of your chickens.

Most predators are opportunistic, can be more hungry at different times of the year, and will prowl at different times of the day or night. Knowing their patterns and life cycle can help you know how to deal with them. You can take several preventive measures to minimize attracting predators to your garden, such as keeping all potential food sources cleaned up and secured—garbage cans, compost areas, even bird seed. Also make sure you have barriers such as strong, proper fencing and secure latches on coops to keep predators away from your flock. Such deterrents can help prevent predator problems, although methods and successes vary greatly.

The Guardian Canine

Dogs are one of the most common chicken predators. There is no denying the strong predator drive in most of our domesticated canine friends when they see a chicken squawking or flapping its wings. Dogs descend from wild hunters with an instinct to kill small animals. Swept up in the ancient thrill of the chase, a dog can quickly snap up and kill a chicken, then drop it later. If you have a problem with a dog you do not know, call the local authorities, whether the sheriff or the local animal control center for answers.

Dogs, on the other hand, can be a great livestock guardians if they are trained properly. Yes, a dog can be a chicken's best friend and help protect the birds from predators. If a puppy is raised with the birds, it can bond with the birds and treat the chickens as a part of its pack, keeping watch over the flock. Just by having the birds and dogs together in the same area will help keep predators away because of the dog's smells.

If your dog was not raised with chickens, you should get it some training to teach it not to follow its instincts to go after its natural prey. Do this right away; don't wait for an unfortunate incident to occur. Even a canine chicken killer can be reformed with some simple training: I know firsthand. My dog was to blame for a few chickens' deaths, and I quickly changed that. It takes a few sessions of training and some patience.

Another thing to know about canine guardians is that certain breeds are better than others for this role. For example, a chihuahua that lives indoors may not be as effective as a Great Pyrenees that has been bred for centuries to watch over flocks of livestock. Other well-known livestock guardian breeds are the old English sheep dog, the border collie, the kuvasz, and Anatolian shepherds. But most working dog breeds will perform well as protectors if trained. A dog that is kept inside most of the day and night may not be very in tune with what goes on outdoors. An outdoor dog with a protective personality, suited for your climate, will be most efficient as a poultry guardian.

OPPOSITE: Malakhi, a pit bull terrier, watches over the garden from the deck.

If your chickens are added to your family after you already have a dog, there are some training steps you should undertake right away. You should repeat these steps several times each, so the dog gets used to the birds and has been conditioned to have good behavior by receiving rewards.

○ **STEP 1:** To introduce a dog to chickens, first let them meet while having a fence or barrier between them so there's no contact. Keep the dog on a leash and under control, since the chickens may be easily stressed with a predator close by. If your dog behaves and can "sit and stay" while the chickens move about in their space, reward the dog with praise or a treat. Use command words that you already use with your dog, such as "gentle" or "leave it," depending on your training style. If the dog is excitable and refuses to listen to your commands, give it a firm correction. A quick yank of the leash (or whatever correction method you use, verbal or physical) will get the dog's attention, and you can try again.

○ **STEP 2:** Once the dog behaves well with the chickens separated by a fence, have another person help by holding a chicken. Choose the most mellow and most handled chicken in your flock. Just as you commanded a behavior in the fenced situation for step 1, if you let the dog see the chicken being held and it can remain in control, it is safe to move on to the next step.

○ **STEP 3:** Have the dog tied up in an area where chickens are allowed to free range. Keep the dog supervised and under command at all times. The dog will watch the chickens going about their business and will eventually get bored and lose interest. The appeal of chasing the birds will wear off.

Coyotes

Coyotes, a common canine predator in rural areas, can also be found in urban environments, sometimes in surprisingly large populations. Coyotes will usually hunt during the dusk and dawn hours and carry off one chicken at a time. They do hunt in packs, so if a few chickens are suddenly missing, coyotes might be to blame. Having a good fence and a dog are your best defenses from coyotes.

Cats

Most domesticated cats are naturally drawn to small birds, making baby chicks a tempting and easy target. But when full-sized chickens reach a mature size, cats rarely go after them. Small breeds such as Bantams can be targeted by our feline friends, but fortunately the small breeds often can fly away. Cats are skilled hunters and naturally control rodent populations, making them a welcome addition to most homesteads.

Large Predators

Because of dwindling populations of large predators in suburban and urban environments, it's unlikely that you would see animals like bobcats, cougars, or bears near your home. If you live in a rural area with these large predators nearby, a strong fence with electric reinforcements is your best defense. These large animals not only pose a threat to our chickens but can be dangerous to our other pets as well as our family. It is best to contact your state's Wildlife Department if these predators become a problem.

OPPOSITE, TOP TO BOTTOM: A Black Star hen seeks cover near tall grass. Beyond the coop and the adjacent vegetable garden, there's a free-range area at Old Goat Farm.

FRIENDS & FOES OF HENS IN THE GARDEN

Raccoons

What appear to be cute, fluffy critters with the notable black eye banding can actually be a major headache and even dangerous for you and your chickens. Raccoons are just about everywhere, and they have become even more of a problem in urban areas where fewer natural food sources are available to them. This predator is intelligent and adapts easily to any situation. They go after both the chickens and their eggs, mostly at night. In the spring when there are new litters, the raccoon mother will often venture out during the daytime in search of food.

Raccoons carry diseases and parasites that can affect chickens, other pets, and even humans. When they feel threatened, they can attack you and your animals. These opportunistic omnivores can be a particular pest for gardeners: they eat nuts, berries, and crops like corn and melons. They are often attracted to garbage cans and food left out over night, both in the coop and for other pets, so keep all food sources cleaned up and secured. Raccoons can easily open gate latches, food containers, and garbage cans, creating a big mess. They are great climbers and can pry open chicken wire to get to your birds, so use hardware cloth or other strong fencing material. They often nest near our homes, under decks or in trees, and are difficult to get rid of with repellants. You may need to call your local animal control agency, which can give you advice and possibly help trap the animals to relocate them.

Opossums are similar to raccoons in population range, habits, size, and ability to inflict damage to crops and animals. They look more like cat-sized rats with a pointed face, and will play dead when threatened.

Birds of Prey

I've always had a great respect for majestic raptors, such as eagles, hawks, and owls. Their keen eyesight, agility, and speed make a hunting raptor an impressive sight to see.

➡ **CAUTION:** All raptors are protected by law. The Migratory Bird Treaty Act makes it illegal to kill these birds.

Owls hunt mostly at night, while hawks hunt by day. Their diet consists of small rodents, birds, and reptiles. The smaller the species of raptor, the less likely it is going to go after full-sized chickens. The larger the bird and its wingspan, the more room it needs to be able to land and take off again, to grab a chicken. Bald eagles, for example, need a landing area that is large and long, because they glide in with their wings open which can reach an impressive 8-foot span. The term "chicken hawks" often refers to three species: the red-tailed hawk, the Cooper hawk, and the sharp-shinned hawk, which are probably the most efficient chicken-hunting raptors. They are smaller than eagles and can dive toward prey at high speeds. They live in diverse habitats in most parts of North America and other continents. To protect your chickens from these hunters, provide dense brush the chickens can take shelter in, or have overhead netting and fencing in the most confined areas.

Foxes

Foxes are agile animals that can run at speeds of 30 miles per hour and can outwit the best of chicken keepers. Foxes are most likely to be a problem in rural environments. They usually hunt at dawn and dusk, but will also come out during the day and night. These sneaky omnivores have excellent hearing and eyesight, making them great hunters. They will usually only take one bird at a time, and can come back time and time again if they are not prevented from doing so. It is best to not allow chickens to free range close to dawn or dusk if there are foxes in your area. Having a guardian dog will help alert you if a fox is nearby.

Mink and Weasels

Mink, weasels, and ferrets, which are in the same family, vary in size and are predators of chickens. They hunt during the day but usually raid the coop at night. They can be the trickiest to keep out of a coop, because their slim form allows them to "weasel their way in" through openings as small as 1 inch. They go after eggs and chickens, often decapitating the birds. Their usual habitat is in rural areas near waterways. Some deterrents can work, such as motion lights and noise from human activity, but the sure way to get rid of them is to trap them. Weasels and mink are excellent for controlling rats, their primary food source, but if they are hungry they will go after your flock.

....

OPPOSITE, TOP TO BOTTOM: The bald eagle is a chicken predator. A red-tailed hawk has just captured its prey.

Snakes

Depending on what region you live in, you may encounter certain snake species, especially types of rattlesnakes like cottonmouth and copperheads that can be chicken predators: they eat eggs, chicks, and sometimes even whole chickens. They are attracted to warmth and can often be found sleeping in a coop under a light. Chickens will eat small snakes, making larger species the biggest problem. Snakes are excellent rat and mice hunters, which usually attracts them to a place. Snakes can slither through small openings in coops and some snakes are very good climbers.

Rats

No matter where you live, you are bound to have some rats in your area, even if you never see them. They go after eggs and young chicks, are vectors of disease and parasites, and can take up residence in your backyard. Rats have been known to eat chickens that roost on the ground in the dark, but that is rare. They consume as much feed as an adult chicken, and they leave their unsanitary droppings everywhere. To keep rats away:

○ *Don't give them access to food.* Keep food in containers that have tight lids and cannot be chewed through, like galvanized metal. Be sure to clean up any feed spills and hang feeders up off the ground, so rats cannot easily eat from the feeder.

○ *Don't give them a place to live.* Keep the coop and surrounding areas clean and tidy. Rats will live underneath rubbish, woodpiles, and will even dig beneath flooring to establish their nests.

My first poorly designed chicken coop was a perfect shelter for rats: they chewed holes to get in, lived underneath the structure, and ate any uneaten food that fell onto the floor. By the time we noticed the problem, we had a healthy rat population under our chicken coop. We cleaned up, remodeled, and the problem was gone.

PREDATOR CONTROL AND DETERRENTS

Predators can cause havoc in the lives of many chicken keepers, and it is difficult to solve the problems. Before taking any action, contact local authorities because some animals are protected by law, and your local animal control agency can inform you of the laws and common solutions in your area. Some of the agencies that may assist you in trapping a problem animal or have good resources are: the sheriff or local police, the state Fish and Wildlife Department, or the Department of Natural Resources.

Trapping

Trapping predators to relocate them is a common way to try to get rid of an animal problem. The local humane society often has traps you can borrow to trap animals like feral cats and raccoons. Using an opened can of cat food or tuna fish or a container of peanut butter will attract most animals into a trap. Be extremely careful about approaching and handling the animal once it is trapped, since they can be dangerous.

Poison

Using poison as a control method can have deadly consequences if ingested by the wrong animal. Dogs, cats, and chickens can easily fall victim to poison intended for pests such as rats and mice. There are many types of poisons. Be sure to place the poison in a bait box that is designed for the rodents, to ensure that pets will not get to it. Also, if you find a dead poisoned rodent, be sure to bury it deeply so it doesn't get eaten.

Deterrents

Using predator deterrents is worth a try to prevent them from coming into our spaces. Predators are intelligent and determined, and will learn over time that some of your deterrents are harmless, so it is important to frequently change your tactics to keep predators guessing. Common deterrents are stimuli disruptors, and can include fake owls or loud and shiny objects such as pie tins. Lighting on motion sensors can be useful for coops prone to attacks at night. For flocks subject to air attacks, you can use tactics like you would use to keep birds away from your garden crops—reflective tape, suspended CDs, or scare balloons. Deterrents that are odor-based have limited success because weather elements shorten their life span and effectiveness. But if the odor is applied regularly, it can be more effective, such as the urine from a predator, such as dogs, or even from humans. Yes, it may seem gross, but it works. Dead animals (coyotes, crows) have been used as deterrents hung on the coop or fence. Farmers today still use this tactic, but it will likely be frowned on by neighbors in urban environments because of sanitation and smell.

OTHER GARDEN FOWL

Chickens are not the only beneficial fowl you can keep in the garden. It is important to know the characteristics and behaviors of other types of poultry before getting new birds.

Guineas

This interesting spotted bird is a little larger than a chicken but is a much different type of fowl. Guineas are nicknamed the "watchdog" in the poultry world, because of their patrolling behavior and ability to sense visitors. When they alert you of a trespasser, their call is loud and lasts for about 20 seconds. This can be a beneficial behavior for the poultry owner but it can be annoying to neighbors.

Guineas are raised for both meat and eggs, and are good in the garden because they don't scratch. They are great insect hunters, and it is common to see them dart across the garden after a moth or grasshopper. They are louder than chickens, range farther than chickens, and fly higher than chickens, but they are basically disease free and have the same if not more interesting personalities.

Guineas can live with chickens, but if they are not trained to roost in a coop, they will sleep in the trees, talking into the hours of the night, making them an easy target for night predators. It is important to establish a routine with guineas, so they are trained to go into a coop at night. Just like acclimating a chicken to your routines, the same must be done with guineas. They are especially great for eating ticks, which is helpful with the growing Lyme disease problem.

Turkeys

Turkeys are large, and I would not recommend letting them free range in a garden setting. Their feet are as big as our hands, and because they can carry up to a whopping 38 pounds, they can do a lot of damage to plants. Even though they are heavy, many breeds can fly. While they don't scratch, they will crush entire plants and topple others. Their height and weight will make them the dominant fowl in the barnyard. I've had a few chickens stand up to turkeys, but they are demanding when it comes to feeding time and will push the chickens out of the way to get to food. I have seen a turkey grab a chicken in the way of its food by the leg and with one swift movement fling the bird several feet. Turkeys are usually friendly and curious but can be intimidating, especially to small children. They are susceptible to blackhead, a disease with a high mortality rate that can be spread from other poultry that are carriers but are more resistant to the disease.

Ducks

Ducks can be beneficial to the landscape by eating slugs and grubs, and they don't scratch. When I first got ducklings from my local feed store, the attendant warned me that they were messy. They are indeed very messy. It is not a good idea to house them with chickens, because they will get water everywhere and damp conditions are not good for a chicken's health. They do not need a large pond, but they do need lots of water to swim in, preferably not stagnant water. I've seen many people use a permanent kiddy pool for their ducks, but you will need to consider the volume of water for the quantity of birds you have and the amount of water needed to keep it clean and fresh.

Ducks are hardy and have few health problems, yet they are sometimes shy and skittish. This potential trait can make them hard to catch and easily stressed. Their personalities are often docile and you might think they would be a good match for chickens, but ducks can be aggressive, especially the males (drakes) who are looking to mate. They need very little housing and can live in a simple shelter similar to a doghouse. Ducks are good for eggs and meat and grow to a processing size quicker than any other fowl. There are many breeds of ducks, some better producers, some quieter, and some better brooders than others. A downside to keeping ducks in the garden is that their flat feet can be destructive to small plants and can cause the soil to be compacted, especially in moist conditions. The Muscovy duck is a popular breed of domesticated ducks for homesteaders; it doesn't quack.

PAGE 189: This hen is standing on creeping Jenny (*Lysimachia nummularia*), a vigorous (and sometimes aggressive) groundcover for shade. OPPOSITE, CLOCKWISE: Female Muscovy ducks forage in the grass. A Muscovy drake (male) duck in the garden. Tom and Jerry, two male Eastern turkeys, strut at the Old Goat Farm.

 **LIVESTOCK**

Most barnyards can have several types of animals living in harmony. But when you keep several types of animals together, there are many things to be aware of.

How you feed the animals and the timing of feeding is probably the most critical to keeping peace and good health on the farm. Food should be kept separate for each type of animal, because it is easy for animals to eat the wrong food and become sick. In my barn, all animals come running when the food is being dished out. I have always fed the largest and most dominant animals first, which usually means the chickens are last to eat. If I were to feed the chickens first, a hungry horse or goat could easily push through the crowd to gobble up the feed.

Separate housing for all animals is important but not always necessary. Chickens can easily co-mingle with other animals, but certain situations are not workable for all animals involved. If chickens share housing with other animals and roost where they shouldn't, chicken manure can get into the bedding of the animals and into their feeders and water, creating a health problem. So provide plenty of roosting bars and covered feeders and water with chickens.

Large Livestock

Horses and cattle can easily injure chickens with a quick step in the wrong direction. I have nursed countless chickens from hoof injuries, and while such an injury is seldom fatal, it can take many weeks to heal. Also, wrapping a bandage on a chicken's foot is not easy to do, and chickens are quick to peck off a bandage.

The manure of large livestock contains undigested feed and seeds that the chickens eat and benefit from. This symbiotic relationship also helps keep flying insects and internal parasites down in the yard, since chickens gobble them up. Chickens also help distribute the piles of manure, exposing them to sun and air, which are the best natural sanitizers. Having chickens around is also helpful in toning down a horse's tendency to spook.

Goats

Goats are unique, and their personalities can range from docile to aggressive or simply overexcitable. They can cause damage and injury to chickens like the large livestock can. I have a rescue goat that has slept with the chickens from day one. They even ride on her back. But chicken manure is said to be dangerous to baby goats. So if goat husbandry is in your plans, avoid any type of co-housing between the two. Another consideration with sharing pasture space with different barnyard animals is that they may be affected by different plants in the pasture. Goats are known to eat just about everything, while other livestock and chickens may not bother with many plants that can be dangerous to goats. Chickens have been known to eat external parasites off sheep, especially after shearing.

Pigs

Pigs are not always a friend to chickens. If the pigs are housed in a specific area, the chickens may wander in because they are attracted to the pig feed. Pigs can and will chase after and eat chickens once they are old enough to understand that chickens taste good. Some farmers may have great luck and find that this hasn't happened for them, but it may not be worth taking that risk.

Bees

The birds and the bees seem to have an agreement. Most chickens are instinctively cautious and don't go after honeybees, but will eat the dead, discarded worker bees around the hive that are a great high-protein food source for the birds. Chickens will eat live wasps and hornets, and on occasion will develop an acquired taste for live honeybees and go after them as they come and go from the hive. If this happens, you can place wire fencing around the hive or keep it up on blocks, out of reach.

OPPOSITE: Panda the pygmy goat and Willow the horse share their hay with chickens in the pasture.

POULTRY DISEASES AND PARASITES

Of the many poultry illnesses and diseases, I will just address some of the most common ailments in backyard chickens. If you want more information on poultry health, consult the Resources section at the back of the book.

It is widely believed that animals know how to self medicate with plants to treat illness and parasites. Zoopharmacognosy is a growing science that collects pharmacological knowledge from the animal kingdom and is a foundation for many herbal health remedies. Many human medicines are derived from plants that we have watched animals use and benefit from.

In the many years I've kept free-range chickens, I have not yet had to deal with disease or damaging levels of parasites. I believe that by having enough room to roam and diverse plants to forage, free-range chickens have healthier immune systems and can eat specific plants to prevent or cure ailments. By allowing chickens to behave as they naturally would, they will have a diverse diet and will choose what they need to survive. Animal species have survived and evolved for centuries without human intervention.

Diseases

You might not encounter poultry disease in a small, well-managed backyard flock. But chickens can be affected by respiratory, viral, and bacterial diseases, many of which are contagious and can spread to other members of the flock. Injury and parasites are also problems with chickens.

Being aware of what a normal, healthy chicken looks like makes an unhealthy chicken easy to spot.

➡ CAUTION: If the bird is alert, has clear eyes, and is active, this is a normal bird. Some symptoms of a sick chicken are watery or mucousy eyes; nasal discharge; panting, wheezing, coughing, sneezing; skin discoloration, either pale or bluish; wartlike lesions; unusual stool—diarrhea, bloody, all white, green, watery; not eating or drinking water; drinking excessively; lameness; tremors; pale combs; and swollen joints or inflamed skin. If you see any of these symptoms, separate the chicken right away, investigate the problem so it can be treated, or take the bird to a vet for diagnosis and care.

common fears and bio-security

Chickens have immune systems and are susceptible to illness and disease just like humans. Birds can become ill from many factors related to their overall health, dirty housing, proximity and health of the other birds in the flock or in the wild, even rodents. Prevention is the best defense in keeping your chickens healthy. Because of recent events covered in the media, people have two main fears about keeping chickens: salmonella and the avian flu. Here are the facts:

○ THE AVIAN FLU. This illness is passed through feces and saliva in water and shared feeders. Humans cannot get avian flu from simply eating chicken eggs or meat, and most large outbreaks have been in large commercial poultry farming operations. Human infection is extremely rare.

○ SALMONELLA. This common bacteria causes illness in humans and can be found in many things we eat. It comes from fecal matter of infected birds or other animals. We've seen recent recalls of salmonella-bearing meat and eggs as well as produce and grains. Most outbreaks are in large commercial poultry or food operations.

➡ IMPORTANT: The chances of getting sick from your backyard flock are slim if you always practice safe food preparation habits and good flock management: Keep your flock healthy—well fed, fresh water, and access to fresh air, sunlight, and greens; keep the coop and nesting boxes clean and rodent free; always wash your hands thoroughly after handling the birds and their manure; collect eggs quickly and discard any that are cracked, broken, or leaking; cook the eggs and meat thoroughly and don't eat them raw; keep a pair of shoes designated for wearing in the coop and sanitize them regularly.

Parasites

Chickens can have internal and external parasites that can potentially harm them. Many chickens live with parasites and have no ill effects because the parasite population never gets out of hand. Parasites can come from many hosts, including rodents, new birds introduced to the flock, or wild birds. The best preventive measure is good flock management.

o *Internal parasites.* Internal parasites that commonly invade chickens include different types of worms and protozoa, and they usually live in the digestive tract but can also invade other parts of their bodies. These internal parasites may be harder to diagnose because the organisms are not visible, but they can cause blood loss, nutrient loss, and sometimes death. Some signs to be aware of are diarrhea or odd-colored stool, lethargy, pale yolks, and hunched-over birds. Wheezing can also be a sign of gapeworms, which infect the lungs and airway. To diagnose the problem, bring a fresh manure sample to your vet. Many people use wormers as a preventive measure, but it is not a one-time fix because the worm life cycle may not be broken if unhatched eggs are still in the chicken's system. Chickens can often cure themselves by browsing on plants with vermifuge properties. If you choose to use chemical dewormers, just be sure to follow directions carefully and be aware that you should throw out any eggs laid while the pesticide moves through the chicken's body.

o *External parasites.* There are several different types of external mites and lice that feed on the blood and feathers of chickens. Most chickens commonly will live with a small population of external parasites at one time or another. Larger parasite populations can make chickens anemic, causing a drop in egg production and sometimes death. Some signs to watch for are excessive scratching or picking at themselves, bare patches of skin, and damaged feathers. Preventive measures include a clean living environment, and you can use branches from trees with insect repellant properties for roosting bars, such as juniper and cedar. Growing plants with insecticidal properties near the coop also may help. Also be sure to provide a dry area for the hens to take dust baths.

diatomaceous earth

This natural material is a mineral dust that comes from diatoms, which are fossilized microscopic shells that can cut through parasites like shards of glass, leaving them dehydrated, which leads to their quick death. This material has been used as a preventive wormer internally in both animals and humans for generations. Externally, it is said to be helpful in pest control, for insects such as lice, fleas, and mites by dusting the nesting boxes, bedding, and adding it to your chickens' dust bath area.

Because of the fine particle size of this material, wear a mask when handling this material. Use caution when applying this material when children and young animals are nearby, because they are more susceptible to lung irritants. Buy food-grade diatomaceous earth, not the diatomaceous earth that is used for pool cleaning and is a known carcinogen.

BELOW: The dust bath, an important grooming behavior, helps prevent external parasites.

TROUBLESHOOTING COMMON CHICKEN PROBLEMS

Chicken owners will encounter several common problems.

Feather Loss

Missing feathers in chickens can be caused by many circumstances, some normal and some not. The situation may need your intervention to prevent infection, cannibalism, or even death. Some common reasons for feather loss appear here; external parasites may also be a factor.

Molting. Molting is a natural process in many animals of shedding their skin, fur, or, in a bird's case, feathers. For chickens, this time of feather replacement can take approximately 4 to 12 weeks. It happens usually once a year in the fall when the days get shorter, and during this time a hen's reproduction system gets a rest and she stops laying eggs. Artificial lights can put off this natural process, making it occur at random times of the year, but every chicken will go through it at some point, usually starting at 18 months. While a chicken is molting, do what you can to reduce stress, since the bird is conserving energy to create new feathers. Also be aware that the social pecking order may change, and a molting chicken may be bullied.

Mating. When a rooster mounts a hen for mating, he stands on top of her back, and she will usually loose a few (if not many) feathers there and at the base of her back and neck. This behavior is normal, and as long as there are no wounds or swelling where the feathers were removed, no action is necessary. You can purchase a mating saddle to prevent injury, and put it on the back of the hen.

Feather picking. Chickens have a complex social hierarchy. For both hens and roosters, they establish a "pecking order" that determines which bird is at the top and which one is at the bottom. Social conflicts can result in chickens having their feathers pecked, leaving the skin exposed, which can lead to cannibalism if the behavior is not stopped. This is most common in chickens that live in overcrowded spaces and don't get adequate food, water, or exercise. If a chicken is free range, it can escape the bullying. Bored chickens also engage in this behavior.

Egg Issues

Eggs are no doubt a fantastic benefit of keeping a flock, but they can cause concern when everything isn't just right.

Weird eggs. Most chicken keepers will see a weird egg from time to time, whether it has a soft shell, is misshapen, or has double yolks. This kind of abnormality is common and is mostly just a fluke, but sometimes it can give you clues about your chicken's health. Nutrient deficiencies are a common cause for soft shells, with calcium often being the missing link as the egg moved through the oviduct. Young pullets will often produce double yolks or missing yolks, even small eggs, which just indicates that their cycle is getting ready to set up shop before the real egg laying begins. Sometimes a disruption in a chicken's environment or new stresses will cause abnormal eggs. Genetics can be the culprit as well. Double shells are rare, and blood specks in the yolk are common. These abnormalities are not much to worry about. But if abnormal eggs become the norm, you may need to take the hen to the vet. If you are concerned about the taste of your eggs or the color of yolk, that is a result of what your chickens are eating. If you find a worm in your egg, that is a sign of a serious infestation of internal parasites.

Egg-bound hens and prolapsed vents. Occasionally an egg can get stuck inside a hen, which can lead to a serious problem. This usually doesn't happen to healthy hens that are fed properly and get lots of exercise. It is more common in older or overweight hens. If it happens once, it is likely to happen again. An egg-bound chicken will stop laying and will act constipated, hanging out in the nest box all day with no egg to show for it. Other signs to look for are tail pumping or just lethargic behavior with fluffed-up feathers. You can try to help her out, but that doesn't always end in happy results. Seek advice from your vet.

OPPOSITE: A weathered metal chicken lives in one of the beautiful gardens at Old Goat Farm where chickens can't range.

Naughty Hens

New chicken owners are often surprised by their hens and can be easily frustrated by behavior they do not understand. Chickens are social and hormonal creatures, and when we have them living in ways that are different from how they would live naturally, they are prone to behaviors that can be damaging to themselves or that are simply normal but just catch us off guard. It is important to know and understand why these behaviors happen, so we know how to best deal with them.

Broody Hens

Natural broody behavior is often seen as a common problem, but it can also be an asset to a breeder. Hormonally, the hen is just doing what her body and instincts tell her to do, which is to sit on the nest to hatch her babies at all costs. Sometimes this behavior can seem threatening, because the hen will fluff up her feathers, make a growling noise, and refuse to leave her nest. Some breeds are known for this behavior more than others. Hens may need to be escorted out of the nesting box to get some exercise and to get water and food if you intend to let her sit and hatch eggs. If this is an undesirable trait, various techniques are recommended, all with limited success. My recommendation is to just let nature takes its course, which can last a few weeks and may not be ideal for everyone. I look at this much like molting: it gives the hen a chance to rest from continuous laying.

Cannibalism

Cannibalistic behavior can happen in any flock, but is usually a result of chickens being either deprived of food, space, or being put in an environment that is unnaturally bright or lacks any stimuli for the chickens. It can innocently start as boredom, and begins with feather picking of other chickens, which can quickly escalate into creating open wounds and eating flesh. This goes beyond chickens establishing a social pecking order, but will most likely happen to the chicken of the lowest ranking, the youngest, or an injured or diseased bird. This behavior is much like egg eating: it can be learned and spread within a flock. Beak trimming is done in commercially run poultry houses to prevent cannibalism, but this is cruel and I don't recommend it for small backyard flocks. To help prevent cannibalism from happening, make sure there is adequate space for all birds and proper diet with lots of room to roam and get away from each other if necessary. If you notice a chicken bully or a chicken being bullied to the point of drawing blood, separate that bird from the flock and let it recover.

Egg Eaters

This chicken behavior is incredibly frustrating for the chicken keeper. Once a chicken tastes an egg, it commonly will relish the flavor and will crack open eggs to eat them on a regular basis. To help prevent this, be sure to gather eggs regularly, provide plenty of bedding in nest boxes, and clean up and remove any cracked or broken eggs quickly. Sometimes it is easy to see who the culprit was if you find egg yolk on the troublemaker's beak, or you might catch her in the act. Isolate the offender before she can teach others this behavior.

Egg eating is often considered a result of a calcium-deficient diet, but if you are already feeding enough crushed oyster shell (or another source of calcium), there are a couple of approaches you can try to stop the behavior. Place a fake, hard egg in the nest box—a golf ball, or a wooden or ceramic egg will work. I have found a real-looking glass egg that I leave in my nest boxes at all times. A few pecks at this solid object and the egg-eating chicken will usually stop the behavior. Another method which is said to work is to use a real egg and poke a small hole in it, blow out the insides, and refill it with mustard or hot sauce. Once the chicken gets through that shell and has a taste of the mustard, that should be the end of the bad habit. And if not, there is always the stew pot.

OPPOSITE: Cheeky, an Easter Egger hen, wanders through a garden.

INJURY IN THE GARDEN

Chickens are prone to injury, but usually are quick to learn to get out of harm's way when allowed to free range. If a chicken has an injury or signs of illness, isolate the bird to keep it from being pecked by its flock mates, and to give you a chance to monitor and treat the bird.

I keep a chicken medical kit handy in case of emergency. I have it in an enclosed storage bin that is kept dry and free of dust. You never know when something might happen, and a trip to the vet can be costly. It can be hard to justify a $200 vet bill for a small wound, but left unattended, an injury can become infected and possibly lead to death.

A few things to keep in mind with acting as chicken doctor in case of an emergency: If there is a need to medicate your birds, be sure to read the labels and follow directions carefully. Many medications can be passed on through eggs and stored in the bird's meat, so it is important to make sure the meat and eggs from a medicated bird are monitored, and disposed of as necessary. Chickens with open wounds should be isolated from the flock.

OPPOSITE: A Rhode Island Red in the garden.

WHEN A CHICKEN DIES

Eventually all chicken keepers will lose a chicken to old age, predators, or illness. A chicken's lifespan is approximately seven years. If the chicken dies from unknown reasons, it is a good idea to send the bird off for a necropsy to determine the cause of death, so that you know whether you are dealing with a contagious disease that might affect your flock. Your avian or poultry veterinarian, local Extension Studies office, or health department can give you information how to deal with this situation, so do your research so you have that information at hand. If you discover that your bird has died from old age or predator attack, it is best to bury it; then its nutrients can feed the soil.

a simple chicken emergency kit

A simple chicken emergency kit should include:

○ **STYPTIC POWDER.** To help blood clot, use cornstarch or flour, or use Kwik Stop which can be found at most pet stores.
○ **ANTIBIOTIC OINTMENT.** Be sure to buy a product intended for animal use.
○ **VET WRAP.** This bandage clings to itself and inhibits bleeding without cutting off circulation.
○ **HYDROGEN PEROXIDE.** Use this to wash wounds.
○ **STERILE GLOVES.**
○ **STERILE COTTON BALLS AND SWABS.** Use these to help clean and dry wounds.
○ **TOWEL.** You can wrap the bird in this for safe handling.
○ **LIST OF VETERINARIAN PHONE NUMBERS** to call in an emergency.

afterword

IN OUR FAMILY, it has become tradition to pick out a few new chicks to add to the flock each spring, to replace older hens or ones that have passed away. Just like the excitement of watching buds break and flowers bloom, these baby birds bring hope and reawakening after a long winter.

Whether you keep chickens for eggs or meat, or treat them like your feathered children, these birds are a blessing to our lives and we should be thankful for what they offer us. They are complex social creatures and we are privileged to give them a good home. Their unique personalities and amusing antics make us laugh and help give us a grounded perspective.

Gardens and chickens can certainly work hand in hand with some forethought and good management. Now that you have the tools to design your own beautiful chicken-friendly garden, you can share your space with a flock and have a happy co-existence in your backyard.

**"Just living is not enough.
One must have sunshine,
freedom, and a little flower."**

—HANS CHRISTIAN ANDERSEN

conversions and hardiness zones

plant hardiness zones: average annual minimum temperature

ZONE	TEMPERATURE (°F)	TEMPERATURE (°C)
1	−50 and below	−46 and below
2	−50 to −40	−46 to −40
3	−40 to −30	−40 to −34
4	−30 to −20	−34 to −29
5	−20 to −10	−29 to −23
6	−10 to 0	−23 to −18
7	0 to 10	−18 to −12
8	10 to 20	−12 to −7
9	20 to 30	−7 to −1
10	30 to 40	−1 to 4
11	40 and above	4 and above

(INCH / M)

1/32 inch = 0.5 mm
1/16 inch = 1.5 mm
1/8 inch = 3.2 mm
1/4 inch = 6.4 mm

(INCH / CM)

1/2 inch = 1.3 cm
3/4 inch = 1.9 cm
1 inch = 2.5 cm
10 inches = 25.4 cm
18 inches = 45.7 cm

(FEET / M)

1 foot = 0.3 m
1 yard / 3 feet = 0.9 m
200 yards = 183 m
600 yards = 549 m
10 sq ft = 0.9 m2
100 sq ft = 9.2 m2
1000 sq ft = 92.9 m2

(MILE / Km)

1 mile = 1.6 km
30 miles = 48.3 km

(POUND / KG)

1 pound = 0.4 kg
10 pounds = 4.5 kg

(ACRE / HECTARE)

1/10 acre = 0.04 ha
1/4 acre = 0.10 ha
1/2 acre = 0.20 ha
1 acre = 0.40 ha

(°F / °C)

1.0 = −17.2
10 = −2.2
50 = 10.0
−30 = −34.4

$°C = 5/9 \times (°F-32)$
$°F = (9/5 \times °C) + 32$

To see the U.S. Department of Agriculture Hardiness Zone Map, go to the U.S. National Arboretum site at: http://www.usna.usda.gov/Hardzone/ushzmap.html.

BOOKS

Damerow, Gail. 1995. *Storey's Guide to Raising Chickens.* North Adams, Massachusetts: Storey Books.

Foreman, Patricia. 2010. *City Chicks.* Buena Vista, Virginia: Good Earth Publications.

Foreman, Patricia, and Andy Lee. 2002. *Day Range Poultry.* Buena Vista, Virginia: Good Earth Publications.

Hemenway, Toby. 2001. *Gaia's Garden: A Guide to Home-Scale Permaculture.* Junction, Vermont: Chelsea Green.

Lee, Andy. 1994. *Chicken Tractor: A Gardeners Guide to Happy Hens and Healthy Soil.* Buena Vista, Virginia: Good Earth Publications.

Ludlow, Rob, and Kimberly Willis. 2009. *Raising Chickens for Dummies.* Hoboken, New Jersey: Wiley.

Madigan, Carleen, Ed. 2009. *The Backyard Homestead.* Adams, Massachusetts: Storey Books.

Moore, Allana. 2007. *Backyard Poultry Naturally: A Complete Guide to Raising Chickens and Ducks Naturally.* Austin, Texas: Acres.

Mollison, Bill. 1991. *Introduction to Permaculture.* Sisters Creek, Tasmania: Tagari Publications.

MAIL-ORDER CHICKS AND SUPPLIES

Dunlap Hatchery
Caldwell, Idaho, 208-459-9088
www.dunlaphatchery.net

Egg Cartons.com
Manchaug, Massachusetts, 888-852-5340
www.eggcartons.com

Lazy 54 Farm
Hubbard, Oregon, 877-344-2050
www.shankshatchery.com

McMurray Hatchery
Webster City, Iowa, 800-456-3280
www.mcmurrayhatchery.com

My Pet Chicken
Norwalk, Connecticut, 888-460-1529
www.mypetchicken.com

P & T Poultry
Mortimer, Shropshire, England, 01584 890263
www.pandtpoultry.co.uk

Egg Cartons.com ...

Rare Poultry Society
www.rarepoultrysociety.co.uk

Rochester Hatchery
Westlock, AB, Canada, 780-307-3622
www.rochesterhatchery.com

Sand Hill Preservation Center
Calamus, Iowa, 563-246-2299
www.sandhillpreservation.com

The Traditional British Fowl Company
www.traditionalbritishfowl.co.uk

WEB RESOURCES: INFORMATION, ORGANIZATIONS, AND MAGAZINES

American College of Poultry Veterinarians
www.acpv.info

American Poultry Association
www.amerpoultryassn.com

Backyard Chickens
www.backyardchickens.com

Backyard Poultry Magazine
www.backyardpoultrymag.com

Canadian Poultry Magazine
www.canadianpoultrymag.com

Domestic Fowl Trust
www.domesticfowltrust.co.uk

Henkeepers Association
www.henkeepersassociation.co.uk

Hobby Farms Magazine
www.hobbyfarms.com

Mother Earth News
www.motherearthnews.com

Permaculture Activist Magazine
www.permacultureactivist.net

Permaculture Magazine
www.permaculture.co.uk

Plants for a Future
www.pfaf.org

The Poultry Club
www.poultryclub.org

Poultry Club of Great Britain
www.poultryclub.org

Poultry Press
www.poultrypress.com

Practical Poultry
www.practicalpoultry.co.uk

USDA Extension Services
www.csrees.usda.gov/Extension

MAIL-ORDER SEEDS AND PLANTS

Burnt Ridge Nursery
Onalaska, Washington, 360-985-2873
www.burntridgenursery.com

Johnny's Seeds
Winslow, Maine, 877-564-6697
www.johnnyseeds.com

Peaceful Valley Farm Supply
Grass Valley, California, 888-784-1722
www.groworganic.com

One Green World Nursery
Molalla, Oregon, 877-353-4028
www.onegreenworld.com

Raintree Nursery
Morton, Washington, 800-391-8892
www.raintreenursery.com

Seeds of Change
Rancho Dominguez, California, 888-762-7333
www.seedsofchange.com

acknowledgments

WITHOUT THE HELP AND SUPPORT OF NUMEROUS PEOPLE, this book would not have been possible. First and foremost, I owe thanks to my family: my husband, Greg, and boys, Noah and Micah, who were so patient, supportive, and always willing to lend a hand with anything and everything. Thanks to Janet Ensley and Valerie Easton for encouragement and help in guiding me in the right direction straight from the beginning. Many thanks to talented Kate Baldwin for her beautiful photos.

I am forever grateful to the gardeners and chicken owners I've interviewed and profiled. Thank you for allowing us to photograph and share your garden so other chicken owners can be inspired: the Zumwalt Family, the Fries family, Gary Mullen and Greg Graves of Old Goat Farm, Alana and Steve Meyer, everyone at the Bullock Permaculture Homestead, especially the Bullock brothers and Dave Boehnlein, Henning Sehmsdorf, and Elizabeth Simpson at the S&S Homestead Biodynamic Farm, Angela Davis, Theresa Loe, Joe McNally and Gabrielle Roesch, Jennifer Carlson, Ken and Rebecca Reid, Meg Brown, Barbara Lycett, Rachael Vitous and Dan Bauer, and Melody Hooper. Gorgeous coops by Heather Bullard at www.heatherbullard.com, Lee Reid at www.better-coopsandgardens.com, Berg Danielson at www.seattlechickencoops.com, and Nicole Starnes Taylor at www.makedesignstudiollc.com. Tom Sanders, thanks for your beautiful photos of raptors. Thanks to Terry Ryan of Legacy Canine Behavior and Training for the chance to experience Poultry in Motion, and to board-certified avian veterinarians Bruce Singbeil and Rocio Crespo, thanks for taking the time to answer my many chicken health questions.

Last but certainly not least, my heartfelt thanks to my editors, especially Juree Sondker of Timber Press, my mother, Susan McKinley, and close friends Erin Dorsey, Katie Merrell, and Elle Trainor, plus my longtime mentor and teacher, Don Marshall, for all the support along the way.

photo credits

Kate Baldwin:
pages 2, 10, 12 below left and right, 15 below left, 24, 36, 38, 40, 41, 43, 54, 62, 64, 65 above right and below left, 73, 77, 85, 86 above, 87, 90, 92, 96, 97 above left and below left and right, 99, 100, 129, 130, 158, 171, 178, 180, 183, 184 below, 189, 190 below, 193, 196, 221, 222

Heather Bullard:
page 176

Berg Danielson:
page 177

Theresa Loe:
page 146, 148

Lee Reid:
page 172

Tom Sanders:
page 186

Olga Shilina / iStockphoto:
page 78 below right

All other photos by **Jessi Bloom**

index

about the author

JESSI BLOOM IS AN AWARD-WINNING LANDSCAPE DESIGNER whose work emphasizes ecological systems, sustainability, and self-sufficiency. She is a certified professional horticulturalist and certified arborist, as well as a long-time chicken owner with a free-ranging flock in her home garden.

Owner of Pacific Northwest–based landscape design-build firm N.W. Bloom–EcoLogical Landscapes, Jessi has been praised as an innovator in sustainable landscape design. Recognition for her work includes awards from the Washington State Department of Ecology, American Horticultural Society, *Pacific Horticulture* magazine, *Sunset*, *425* magazine, Washington State Nursery and Landscape Association, Washington Association of Landscape Professionals, and the Northwest Flower and Garden Show, including gold medals and the People's Choice award.

Published in 2012 by Timber Press, Inc.

The Haseltine Building
133 S.W. Second Avenue, Suite 450
Portland, Oregon 97204-3527
timberpress.com

2 The Quadrant
135 Salusbury Road
London NW6 6RJ
timberpress.co.uk

Designed by Cat Grishaver

Printed in China

Library of Congress Cataloging-in-Publication Data

Bloom, Jessi.
Free-range chicken gardens: how to create a beautiful, chicken-friendly
yard/Jessi Bloom; with photographs by Kate Baldwin.—1st ed.
    p. cm.
Includes index.
ISBN 978-1-60469-237-2
1. Chickens. 2. Chickens—Housing. 3. Gardening to attract birds. I. Title.
SF487.B64 2012
636.5—dc23

                              2011025180

A catalog record for this book is also available from the British Library.

. . . .

FRONTISPIECE: A Buff Orpington free ranges in Alana Meyer's garden. PAGE 4: A Buckeye hen
peeks out of the shrubbery in Alana Meyer's garden. PAGE 6: A Buff Orpington hen watches an
insect fly alongside this flagstone pathway. PAGE 9: An Australorp hen and her friend forage in
a mixed shrub border. PAGE 211: A chicken made of repurposed materials greets visitors to the
Fries garden. PAGE 219: The ladies at work in Jessi's garden. THIS PAGE: An Ohio Buckeye chicken.